Advances in Building Energy Research

Advances in Building Energy Research

Volume 1

Editor-in-Chief Mat Santamouris

earthscan
from Routledge

First published 2007 by Earthscan

2 Park Square, Milton Park, Abingdon, Oxfordshire OX14 4RN
52 Vanderbilt Avenue, New York, NY 10017

Routledge is an imprint of the Taylor & Francis Group, an informa business

First issued in hardback 2019

ISBN   9781844077526   (hbk)

ISBN   9780367577704   (pbk)

Typeset by Domex e-Data, India
Cover design by Giles Smith

A catalogue record for this book is available from the British Library

Library of Congress Cataloging-in-Publication Data

Advances in building energy research / editor-in-chief, Mat Santamouris.
    v. cm.
Includes bibliographical references.
ISBN-13: 978-1-84407-389-4 (hardback)
ISBN-10: 1-84407-389-0 (hardback)
1. Buildings–Energy conservation. I. Santamouris, M. (Matheos),
1956-TJ163.5.B84A285 2007
696–dc22

                          2007004087

Contents

List of Figures and Tables

FIGURES

TABLES

List of Acronyms and Abbreviations

AC	air-conditioned
ADHRIA	Automated High Dynamic Range Imaging Acquisition
AEB	auto-exposure bracketing
AHU	air handling unit
AI	artificial intelligence
ANFIS	adaptive neural fuzzy inference system
ANN	artificial neural networks
BBRI	Belgian Building Research Institute
BEMS	building energy management system
BLHI	boundary layer heat island
BMS	building management system
BREEAM	BRE's Environmental Assessment Method
BSI	British Standards Institution
BUS	Building Use Studies
CBE	Center for the Built Environment
CCC	Confederation of Construction Clients
CCD	charge coupled device
CDD	cooling degree day
CF	concentrated flow
CFC	chlorofluorocarbon
CFD	computational fluid dynamics
CGF	conventional glass façade
CHP	combined heat and power
CIBSE	Chartered Institute of Building Services Engineers
CLC	CORINE Land Cover
CLHI	canopy layer heat island
CMOS	complementary metal oxide semiconductor
COP	coefficient of performance
CRI	compositional rule of inference
CWEC	Canadian Weather for Energy Calculations
DAI	distributed artifical intelligence
DF	diffuse flow
DGI	daylight glare index
DGP	daylight glare probability
DI	discomfort index
DID	degree of individual dissatisfaction
DP	differential pressure
DSF	double skin façade
EEM	energy efficiency measure
EPBD	Energy Performance of Buildings Directive

ET	effective temperature
EV	exposure value
EXIF	exchangeable image file format
FCU	fan coil unit
FFSI	functioning fuzzy-subset inference
FLC	fuzzy logic controller
F-PID	fuzzy proportional integral derivative
FSC	fuzzy satisfactory cluster
FTC	fault-tolerant supervisory control
GA	genetic algorithms
GIS	geographic information system
GP	genetic programming
GWP	global warming potential
HC	hydrocarbon
HCFC	hydrochlorofluorocarbon
HDD	heating degree day
HDH	heating degree hour
HDRI	high dynamic range imaging
HFC	hydrofluorocarbon
HOPE	Health Optimisation Protocol for Energy-efficient Buildings
HVAC	heating, ventilation and air conditioning
HY	historical year
IAQ	indoor air quality
IEQ	indoor environmental quality
IHEE	infiltration heat exchange efficiency
IWEC	International Weather for Energy Calculations
LCD	liquid crystal display
LDR	low dynamic range
LED	light-emitting diode
LEE	least enthalpy estimator
LEED	Leadership in Energy and Environmental Design
LST	land surface temperature
MEC	manual exposure correction
MMPC	multiple model predictive control
MUHI	micro-urban heat island
NDVI	normalized difference vegetation index
NN	neural network
NV	naturally ventilated
ODP	ozone depletion potential
PD	proportional derivative
PDA	personal digital assitant
PI	proportional integral
PID	proportional integral derivative
PMV	predicted mean vote

PPD	percentage of people dissatisfied
PROBE	Post-occupancy Review of Buildings and their Engineering
PSD	proportional sum derivative
PSF	point spread function
PV	photovoltaic
RAC	room air-conditioner
RBF	radial basis function
RES	renewable energy sources
REV	representative elementary volume
RMSE	root mean square error
SBS	sick building syndrome
SCATs	Smart Controls and Thermal Comfort
s.d.	standard deviation
SDK	software development kit
s.d.m.	standard deviation mean
SET	standard effective temperature
SUHI	surface urban heat island
SVD	singular value decomposition
SWOT	strengths, weaknesses, opportunities and threats
SY	synthetic year
TCL	thermal comfort level
TMM	typical meteorological month
TMY	typical meteorological year
TRDTS	testing room dynamic thermal system
TRY	test reference year
TSV	thermal sensation vote
TWY	typical weather year
TY	typical year
UGR	unified glare rating
UHI	urban heat island
USB	universal serial bus
VDSF	ventilated double skin façade
VOC	volatile organic compound
WYEC	Weather Year for Energy Calculations

On the Typology, Costs, Energy Performance, Environmental Quality and Operational Characteristics of Double Skin Façades in European Buildings

Wolfgang Streicher, Richard Heimrath, Herwig Hengsberger, Thomas Mach, Reinhard Waldner, Gilles Flamant, Xavier Loncour, Gérard Guarracino, Hans Erhorn, Heike Erhorn-Kluttig, Matheos Santamouris, Ifigenia Farou, S. Zerefos, M. Assimakopoulos, Rogério Duarte, Åke Blomsterberg, Lars Sjöberg and Christer Blomquist

Abstract

The project BESTFAÇADE, sponsored by the Energy Intelligent Europe programme of the European Union, and led by MCE-Anlagenbau, Austria, accumulated the state of the art of double skin façades (DSFs) in seven European countries (Austria, Belgium, France, Germany, Greece, Portugal and Sweden). Twenty-eight façades of different buildings in all partner countries of BESTFAÇADE have been analysed for the aspects, types of façade in different countries, DSFs in different climatic regions of Europe, existing simulations and measurements, thermal behaviour, indoor air quality, comfort, user acceptance, energy demand and consumptions, control strategies, integrated building technology, cost (investment, maintenance and operation), resource conservation, environmental impact, comparison to conventional glass façades (CGFs), integration of renewable energy sources into DSFs, as well as non-energy related issues, such as, acoustics, aesthetics, fire protection, moisture, corrosion, durability, maintenance and repair. Most of the buildings are office buildings, followed by schools and service buildings. Nearly all of the buildings have mechanical ventilation systems, and both heating and cooling are performed mostly by air heating/cooling systems. The types of façades are mainly multi-storey and corridor types; in Belgium juxtaposed modules are frequently used. The façade gaps are mostly naturally ventilated (except for Belgium, where the indoor air is led by mechanical ventilation via the gap to the

> centralized air handling unit). The shading is performed mainly with Venetian blinds located in the gap. Unfortunately data on energy demand and temperatures are infrequently measured and rarely available. The cost of DSFs is significantly higher than conventional façades.

■ *Keywords* – double skin façade; typology; technology; costs; performance

INTRODUCTION

Innovative façade concepts are today more relevant than ever. The demand for natural ventilation in commercial buildings is increasing due to growing environmental consciousness while, at the same time, energy consumption for buildings has to be reduced. An advanced façade should allow for a comfortable indoor climate, sound protection and good lighting, while minimizing the demand for auxiliary energy input. Double skin façades (DSFs) have become an important and increasingly used architectural element in office buildings over the last 15 years. They can provide a thermal buffer zone, solar preheating of ventilation air, energy saving, sound, wind and pollutant protection with open windows, night cooling, protection of shading devices, space for energy gaining devices, such as, photovoltaic (PV) cells, and differentiated aesthetical qualities, which is often the main argument in their favour.

Commercial and office buildings with integrated DSFs can be energy efficient buildings providing all the qualities listed above. However, not all DSFs built in recent years perform well. Far from it, in most cases large air-conditioning systems have to compensate for summer overheating problems and energy consumption often exceeds the intended heating energy savings. Therefore this architectural trend has, in many cases, resulted in a step backwards regarding energy efficiency and the possible use of passive solar energy.

The project BESTFAÇADE, sponsored by the Energy Intelligent Europe programme of the European Union, and led by MCE-Anlagenbau, Austria, accumulated the state of the art of double skin façades in seven European countries (Austria, Belgium, France, Germany, Greece, Portugal and Sweden). Twenty-eight façades of different buildings in all partner countries of BESTFAÇADE have been studied by means of a standardized questionnaire. The questionnaire comprises data on location, information about the building and the façade, construction and airflow in the façade, as well as maintenance and cost.

The analysis has been drawn for the aspects, types of façade in different countries, DSFs in different climatic regions of Europe, existing simulations and measurements, thermal behaviour, indoor air quality, comfort, user acceptance, energy demand and consumptions, control strategies, integrated building technology, cost (investment, maintenance and operation), resource conservation, environmental impact, comparison to CGFs, integration of renewable energy sources into DSF, as well as non-energy related issues, such as, acoustics, aesthetics, fire protection, moisture, corrosion, durability, maintenance and repair.

DEFINITION

The DSF concept has been specified in a number of definitions:

A double skin façade can be defined as a traditional single façade doubled inside or outside by a second, essentially glazed façade. Each of these two façades is commonly called a skin. A ventilated cavity – having a width which can range from several centimetres to several metres – is located between these two skins. Automated equipment, such as shading devices, motorized openings or fans, are most often integrated into the façade. The main difference between a ventilated double façade and an airtight multiple glazing, whether or not integrating a shading device in the cavity separating the glazing, lies in the intentional and possibly controlled ventilation of the cavity of the double façade. (BBRI, 2004)

[DSFs are] essentially a pair of glass 'skins' separated by an air corridor. The main layer of glass is usually insulating. The air space between the layers of glass acts as insulation against temperature extremes, winds, and sound. Sun-shading devices are often located between the two skins. All elements can be arranged differently into numbers of permutations and combinations of both solid and diaphanous membranes. (Harrison and Meyer-Boake, 2003)

The Double Skin Façade is a system consisting of two glass skins placed in such a way that air flows in the intermediate cavity. The ventilation of the cavity can be natural, fan supported or mechanical. Apart from the type of the ventilation inside the cavity, the origin and destination of the air can differ depending mostly on climatic conditions, the use, the location, the occupational hours of the building and the HVAC [heating, ventilation and air conditioning] strategy. The glass skins can be single or double glazing units with a distance from 20cm up to 2 meters. Often, for protection and heat extraction reasons during the cooling period, solar shading devices are placed inside the cavity. Poirazis (2004)

HISTORY

The history of DSFs has been described in several books, reports and articles. Saelens (2002) mentions that 'in 1849, Jean-Baptiste Jobard, at that time director of the Industrial Museum in Brussels, described an early version of a mechanically ventilated multiple skin façade. He mentions how in winter hot air should be circulated between two glazings, while in summer it should be cold air'. Crespo (1999) and Neubert (1999) claim that, the first instance of a double skin curtain wall appears in 1903 in the Steiff factory in Giengen/Brenz near Ulm, Germany (Figure 1.1). According to them, the priorities were to maximize daylighting while taking into account the cold weather and the strong winds of the region. The solution was a three-storey structure with a ground floor for storage space and two upper floors used for work areas. The structure of the building proved to be successful and two additions were built in 1904 and 1908 with the same double skin system, but using5timber instead of steel in the structure for budget reasons. All buildings are still in use.

Source: Neubert (1999)

FIGURE 1.1 Steiff factory, Giengen/Brenz, Germany

In 1903 Otto Wagner won the competition for the Post Office Savings Bank in Vienna, Austria. The building, built in two phases from 1904 to 1912, has a double skin skylight in the main hall. At the end of the 1920s double skins were being developed with other priorities in mind. Two cases can be clearly identified. In Russia, Moisei Ginzburg experimented with double skin strips in the communal housing blocks of his Narkomfin building (1928) and Le Corbusier designed the Centrosoyus in Moscow. A year later, Le Corbusier started the design for the Cite de Refuge (1929) and the Immeuble Clarte (1930) in Paris and postulated two new features. First, *'la respiration exacte'* ('... an exactly regulated mechanical ventilation system'), and second, *'le mur neutralisant'* ('... neutralizing walls are made of glass or stone or both of them. They consist of two membranes which form a gap of a few centimetres. Through this gap, which envelops the whole building, air is conducted (hot air in Moscow and cold in Dakar). By that the inner surface maintains a constant temperature of 18°C. The building is hermetically sealed! In the future no dust will find its way into the rooms. No flies, no gnats will enter. And no noise!') (Le Corbusier, 1964).

Little or no progress was made in double skin glass construction until the late 1970s and early 1980s. During the 1980s this type of façade started gaining momentum. Most were designed taking into account environmental concerns, like the offices of Leslie and Godwin. In other cases the aesthetic effect of the multiple layers of glass was the principal concern. In the 1990s two factors strongly influenced the proliferation of DSFs. Environmental concerns started influencing architectural design both from a technical standpoint and as a political influence that made 'green buildings' a good image for corporate architecture (Poirazis, 2004).

TECHNICAL DESCRIPTION
FAÇADE CONSTRUCTION

The choice of the glass type for the interior and exterior panes depends on the typology of the façade. In the case of a façade ventilated with outdoor air, an insulating pane (a thermal break) is usually placed at the interior and a single glazing at the exterior side.

However, in the case of a façade ventilated with indoor air, the insulating pane is usually placed at the exterior while the single glazing is at the interior side. For some specific types of façade, the internal glass pane can be opened by the user to allow natural ventilation of the building.

The ventilation of the cavity may be totally natural, fan supported (hybrid) or totally mechanical. The width of the cavity can vary as a function of the applied concept between 10cm to more than 2m. The width influences the physical properties of the façade and also the way that the façade is maintained. If cleaning of all façade panes is not possible from inside and outside of the building, the width of the cavity has to be about 80cm to allow cleaning personal to access the gap. In this case the airflow in the gap has less flow resistance and can therefore be higher compared to a narrow gap. However, more rented or sold space of the building is lost.

Shading devices can be placed inside the cavity for protection. Often venetian blinds can be used. The characteristics and position of the blind influence the physical behaviour of the cavity because the blind absorbs and reflects energy from radiation. Thus, the selection of the shading device should be made after considering the proper combination between the pane type, the cavity geometry and the ventilation strategy and has a high impact on the daylight situation within the rooms behind. Openings in the external and internal skin and sometimes ventilators allow the ventilation of the cavity. The choice of the proper pane type and shading device is crucial for the function of the DSF system. Different panes can influence the air temperature and thus the flow in the case of a naturally ventilated cavity.

The geometry (mainly width and height of the cavity) and the properties of the blinds (absorbance, reflection and transmission) may also affect the type of airflow in the cavity. When designing a DSF it is important to determine the type, size and position of interior and exterior openings of the cavity since these parameters influence the type of airflow and the air velocity and thus the temperatures in the cavity (more important in high-rise buildings). The design of the interior and exterior openings is also crucial for the indoor flow and consequently the ventilation rate and the thermal comfort of the occupants.

It is really important to understand the performance of the DSF system by studying the physics of the cavity. The geometry of the façade, the choice of the glass panes and shading devices and the size and position of the interior and exterior openings determine the use of the DSF and the heating, ventilation and air-conditioning (HVAC) strategy that has to be followed in order to succeed in improving the indoor environment and reducing energy use. The individual façade design and proper façade integration are key to high building performance.

Compared to traditional office buildings, especially with large glazed façades, office buildings with DSFs can have the following potential advantages:

● Individual window ventilation is almost independent of wind and weather conditions, mainly during sunny winter days and the intermediate season (spring and autumn).
● Heating demand is reduced thanks to preheating of outdoor air.
● Night cooling of the building by opening the inner windows is possible if the façade is well ventilated.
● Security is improved thanks to the two glazed skins.

- There is better sound proofing from external noise sources, for example, at locations with heavy traffic, mainly during window ventilation.
- Exterior (intermediate) solar shading is more efficient, as the shading can also be used during windy days.
- There is potential for the protection of existing façades in renovation applications.

Potential problems with office buildings with DSFs can be:

- poorer cross-ventilation and insufficient removal of heat from the offices rooms during windless periods, when ventilation is mainly provided by natural ventilation;
- hot summer/spring/autumn days leading to high temperatures in office rooms as a result of window ventilation;
- higher investment cost;
- reduced office floor area;
- risk of sound transmission via the façade cavity from one office to another with open windows
- additional cost due to cleaning;
- overestimated energy saving potential;
- more difficult fire protection, depending on the type of façade;
- loss of daylight and increase electric loads by lighting; and
- loss of possible cooling by window ventilation during spring and autumn (low outdoor temperatures).

DOUBLE SKIN FAÇADE BUILDINGS AND HVAC

There are two different approaches. First, a building with its own separate heating, cooling and ventilating system, where a second skin is added to the façade; the cavity of the double skin façade is only ventilated from the outside and is built to reduce noise, house solar shading and light redirection devices. Second, a building where the heating, cooling and ventilating system of the building is integrated into the DSF, for example, by ventilating the building using the cavity of the DSF.

The second alternative is often the most cost-effective option. The risk of the first alternative is having a building with a complete conventional HVAC system surcharged with the added cost of an expensive façade. According to different investigations and technical reports there are some technical benefits with a DSF, benefits which have an impact on the HVAC system:

- All types of DSF offer a protected place within the air gap to mount solar shading and daylight enhancing devices, which then can be used whenever necessary and thereby reduce the cooling load.
- One of the advantages DSF systems may have is that they allow natural (or fan supported) ventilation during some periods of the year, which will reduce the use of electricity for ventilation.
- In winter the cavity forms a thermal buffer zone that reduces heat losses and enables passive thermal gain from solar radiation, which will reduce the heating load.

- DSFs may enable natural ventilation and night-time cooling of the building's thermal mass, which will reduce the use of electricity for ventilation and the cooling load.
- DSFs reduce noise from motor traffic, enabling natural ventilation without noise problems.

EVALUATION OF VARIOUS FAÇADES
LOCATION OF DSF BUILDINGS AND TYPOLOGY IN DIFFERENT COUNTRIES

Table 1.1 and Figure 1.2 show the locations the 28 façades of different buildings in all partner countries of BESTFAÇADE, which have been studied by means of a standardized questionnaire. The questionnaire comprises data on location, information about the building and the façade, construction and route of airflow in the façade, as well as maintenance and cost.

In Austria, the aim was to cover as many as possible different sizes, types and utilizations of buildings with DSFs, for example, newly built DSFs as well as retrofitted ones, offices as well as schools and museums. But unfortunately the smallest, the largest and the most extraordinary DSFs could not be researched, although the managers of these buildings showed high interest in joining the project at the beginning. The example of the small DSF is just two storeys high and is the retrofitting is of three façades of the control room of the fire station in Graz. The main purposes were to improve noise protection and thermal efficiency, and both aims are said to have been achieved by the attached single pane façade with venetian blinds inside the gap. The building that would have been one of the largest researched buildings in the project, the 24-storey Uniqa Tower in Vienna, is said to be one of the most interesting towers among the aspiring high-rise buildings in the city because of its HVAC concept and the good performance of its DSF. There will probably be a chance to get data from this tower in the near future. The third interesting building that

Source: Streicher (2005)

FIGURE 1.2 Buildings analysed within the BESTFAÇADE project

TABLE 1.1 DSF buildings analysed within the BESTFAÇADE project

NO.	COUNTRY	PARTNER	FACADE			
			NAME	CITY	ORIENTATION	UTILIZATION
1			BiSoP	Baden	S / N	school
2			Felbermayr	Salzburg	S	office - n.p.
3	Austria	IWT	Fachhochschule	Kufstein	NW	school / office - n.p.
4			Justizzentrum	Leoben	SE	office - p.
5			Schubertstrasse	Graz	SE	office - n.p.
6			Aula Magna	Louvain-La-Neuve	SE	other
7	Belgium	BBRI	Sony	Zaventem	NE / SW	office - n.p.
8			UCB Center	Brussels	NE / SW	office - n.p.
9			Cité	Lyon	NE	office - n.p.
10	France	LASH-DGCB	EAL	Vaulx en Velin	NE	school
11			Thiers	Lyon	E	office - n.p.
12			Münchner Tor	Munich	N / S / E / W	office - n.p.
13	Germany	IBP	Geschäftsgeb. Süd 1+4	Munich	N / S / E / W	office - n.p.
14			Zentralbibliothek	Ulm	N / S / E / W	library
15			A-A Holdings	Athens	E	office - n.p.
16	Greece	NKUA	Alumil M5	Kilkis-Stavrochori	E	office - n.p.
17			AVAX	Athens	E	office - n.p.
18			CGD	Lisbon	S	office - n.p.
19			Atrium Saldanha	Lisbon	SW	office - n.p. / services
20	Portugal	ISQ	ES Viagens / expo 98	Lisbon	SE	office - n.p. / services
21			Palacio Sotto Mayor	Lisbon	SE	sevices
22			Torre Zen	Lisbon	S	office - n.p. / services
23			ABB	Sollentuna / Stockholm	W	office - n.p.
24			Arlanda	Stockholm	N / S / E / W	other (airport terminal)
25	Sweden	WSP	Glashuset	Stockholm	S	office - p. / school
26			Kista	Kista / Stockholm	S / W	office - n.p.
27			Polishuset	Stockholm	S / W	office - n.p.
28	Germany	IBP	VERU	Holzkirchen	W	test facility

Note: p = public; n.p. = non-public
Source: Streicher (2005)

should have been covered is the 'Kunsthaus Bregenz', which is well known for its architecture. Since the walls of this museum have to be opaque for presentation reasons, the DSF is used to provide daylight for special light ceilings in each storey.

Besides the buildings described above, a special type was covered in the analysis too. In the façade of BiSoP, Baden, the operable windows are bypassing the gap. This seems to be a good compromise for using the interesting aesthetics of the DSF and at the same time avoid many disadvantages such as overheating, condensation and sound transmission. However, natural ventilation by opening of windows is limited by the height of the building because of the increasing wind pressure on the façade.

In Belgium there is a specific situation concerning the concepts of ventilated double skin façades (VDSFs). Indeed, a national project has shown that the majority of VDSFs use an industrialized façade concept where the façade is partitioned per storey with juxtaposed modules and characterized by a single ventilation mode: the indoor air curtain. The façade is used to extract the air from the room with which it is in contact (indoor air curtain). Usually, for the majority of buildings, not only the VDSF but also the HVAC equipment are of the same kind (Figure 1.3).

In Portugal DSF buildings are located mainly in Lisbon, where different architects have designed several that are high-rise. These are mainly privately owned office buildings, some of them belonging to important Portuguese financial institutions. In fact, DSFs were already being designed in Portugal in the 1980s (Caixa Geral de Depósitos, Av. da República), and currently different typologies coexist in the city of Lisbon. These buildings usually have more than five storeys and the most common typologies are corridor façades and multi-storey façades. Aesthetics and energy conservation are some of the main reasons that architects use to support the use of DSFs. Despite the significant number of DSF buildings in Lisbon, and according to the information gathered, until now no comprehensive energetic, acoustic, lighting and user acceptance study of Portuguese DSFs has been made.

In Sweden the interest among architects in applying the technique of double skin glass façades, mainly in new construction of office buildings, has increased over recent years. Such buildings have been built primarily in the Stockholm area, for example, the Kista Science Tower, the ABB-house, the new police house, Glashusett and the Arlanda Terminal F, but there are also examples in other Scandinavian countries. In total there are about ten modern glazed office buildings with DSFs in Sweden. In these cases the purpose of the double skin has been to reduce high indoor temperatures with protected efficient exterior solar shading during

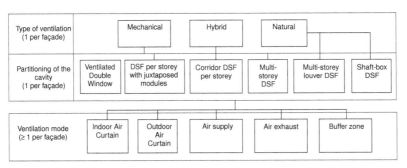

Source: Streicher (2005)

FIGURE 1.3 The typical Belgium DSF within the multitude of varieties

summer, reduce transmission losses during winter, and, in some cases, also to reduce noise from motor traffic. The DSF in Scandinavia has rarely been used for ventilation of the building behind. Modern office buildings in Sweden have high energy savings potential and potential for indoor climate improvements. They may have a lower energy use for heating, but, by contrast, they often have a higher use of electricity than older office buildings. Why are offices with fully glazed façades being built in Sweden? Architecturally an airy, transparent and light building is created, with more access to daylight than in a more traditional office building (Svensson and Åqvist, 2001). Technically it is possible to have protected 'exterior' movable efficient solar shading, to reduce noise from motor traffic and to open windows for ventilation during part of the year (Carlsson, 2003). Swedish buildings with DSFs share many of the characteristics of DSFs in Germany, that is, they are mainly for high-profile, high-quality office buildings (new constructions) and are used when building envelopes with transparency and lightness are wanted and daylight and aesthetics are important.

Examples from Germany are two office buildings in Munich: a major public library in Ulm in the extraordinary shape of a pyramid, and the VERU test facility at the Fraunhofer Institute for Building Physics in Holzkirchen near Munich. Data for the buildings are based on the energy performance certification according to the new standard DIN 18599. During the planning phase of the library in Ulm scientific support was given, including energy performance calculations according to the former 'Wärmeschutzverordnung' and thermal simulations. Detailed energy consumption data are not available for the library, but the total energy consumption levels are known. Data for the VERU test facility are detailed and calculated with the DIN 18599. However, this building is not occupied by users, therefore a user investigation is not possible.

DSF examples from Greece are three office buildings, a hotel that is under renovation and a retail building that is currently under construction. The majority of the DSF buildings are located in Athens, apart from one office building that is located in Kilkis, a northern area of Greece. Different DSF typologies are used in these buildings: the corridor type, the double window, the multi-storey façade and the multi-storey louvre façade. The double window façade was used for acoustic reasons in the hotel building, which is located in a densely built-up area of central Athens with high traffic and noise levels. The other types of DSFs were chosen mainly for aesthetics and energy conservation reasons.

However, in Greece where the climatic conditions encourage the use of natural ventilation and necessitate the control of solar gains in order to prevent overheating, the preferred types of DSF are the multi-storey façade and the multi-storey louvre façade that combine external shading systems and natural ventilation.

FAÇADE TYPES IN DIFFERENT CLIMATIC REGIONS

As buildings with DSFs are mainly erected in the big cities that have a special 'city climate', similar climatic conditions apply for the buildings in all the countries. Thus, although Austria covers different climatic regions such as the pannonian and alpine climates, DSFs are concentrated in the capital city of Vienna. However, DSFs can be found in smaller cities too, for example in Graz, Leoben, Salzburg and Bregenz, but again these share similar climatic conditions. Other countries such as Belgium are too small to have distinctive climatic regions, while in Sweden and Portugal all the researched buildings are situated in the capitals. Therefore, it is not easy to identify special types of DSF specific to a certain region of a country and so climate does not impact on the choice of DSF type applied. Even in Germany there is little point in dividing the country into different climate regions as the differences between them would be minor. However, there are of course small differences concerning solar radiation and temperature (for example, the Freiburg region and the Rhine area have more sunshine and higher temperatures in summer than, for example, the regions near the North Sea).

To structure the BESTFAÇADE results according to climatic conditions, three main regions are proposed:

- the Nordic region, with Sweden as its only representative;
- the moderate region, with Austria, Belgium, France and Germany; and
- the Mediterranean region, with Greece and Portugal.

The following analysis takes these regional divisions and their special circumstances into account. For example, Greece is located in the south-eastern part of Europe between the latitudes of 34° and 42° N, with a meridian extent from 19° to 28° E. The climate in Greece is typical of the Mediterranean climate: mild and rainy winters, relatively warm and dry summers and, generally, extended periods of sunshine throughout most of the year. The year can be subdivided into two main seasons. First, the cold and rainy period lasts from mid-October until the end of March. The coldest months are January and February, with a mean minimum temperature of 5–10°C near the coasts, 0–5°C over the mainland, and lower values over the northern part of the country. Second is the warm and dry season that lasts from April until September. During this period the weather is usually stable, the sky is clear, the sun is bright and there is generally no rainfall. The warmest period occurs during the last ten days of July and the first ten days of August, with a mean maximum temperature of 29–35°C. For the climate of Greece, control of solar gains in the building design is important during the summer periods. Therefore DSFs may lead to overheating during the summer months if there is no appropriate façade design, ventilation technique, building orientation and provision of shading. The climate of Greece encourages the use of natural ventilation in office buildings; however, over recent decades there has been an increased use of air conditioning due to high ambient air temperatures and high internal temperature gains in large office buildings.

Many of the above mentioned potential advantages of office buildings with DSFs are likely to be valid for Sweden as well. In addition, there are other potential problems (Blomsterberg, 2003):

- During warm summer, spring and autumn days high temperatures in office rooms can occur as a result of window ventilation.
- The low altitude of the sun results in fairly high cooling demands during spring and autumn.
- The is a possible risk of high energy use.
- There is a risk of low daylight levels in the central parts of the building, mainly for deep buildings.
- Operation and maintenance costs can be high.

Many modern Swedish office buildings have large glazed façades and some have the additional façade given by a DSF. The simplest and most common system solutions in Sweden entail that the façade is only ventilated to the outside. This usually means that the office building behind has a traditional heating, cooling and ventilation system. Window ventilation is usually not possible, apart from French doors, whose purpose is to gain access to the double skin façade cavity for maintenance.

TYPE OF BUILDINGS AND FAÇADES ANALYSED IN THE BESTFAÇADE PROJECT

Most of the buildings analysed were non-public office buildings, followed by public schools and services (Figure 1.4). None of the buildings were equipped with a DSF in a renovation process and there were no clear main orientation of the façades, which were mainly an architectural element (Figure 1.5). Most of the façades used natural ventilation

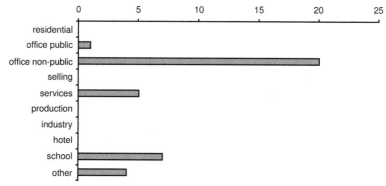

Source: Streicher (2005)

FIGURE 1.4 Utilization of BESTFAÇADE buildings

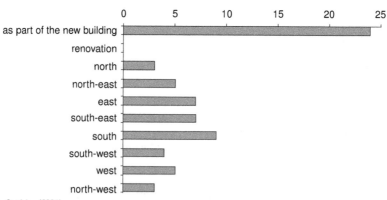

Source: Streicher (2005)

FIGURE 1.5 Implementation and orientation of façades within BESTFAÇADE

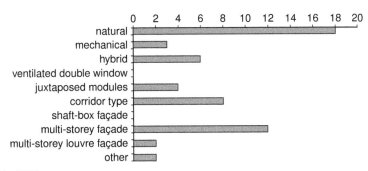

Source: Streicher (2005)

FIGURE 1.6 Type of ventilation and partitioning of the gap

and some a hybrid ventilation scheme. The multi-storey approach was mainly used, while others adopted the corridor type and juxtaposed modules (Figure 1.6.).

EXISTING SIMULATIONS AND MEASUREMENTS
In Austria not many measurements have been done from which data are available. From BiSoP, Baden, south façade intensive measurement data compared to simulation data are available. In this case the aim was to research the physical behaviour of the façade and not primarily its influence on the rooms behind. FH Kufstein has done some simulations and measurements as well. From Felbermayr, Salzburg, some single measurement data are available, while Uniqa has done much work on this but it has not been possible to obtain those data.

In Belgium BBRI has carried out several measurements on DSFs: some *in situ*, some in outdoor test cells and some in laboratories. Different fields were examined: energy, ventilation, acoustics and daylight. A detailed monitoring of the most common concept of DSF applied in Belgium was performed in 2005 in order to determine the thermal and solar properties (in winter and summer) of this kind of façade. Some universities have also performed measurements in laboratories or *in situ*. All these measurements have been realized at the level of the façade component (and not at the level of the building). BBRI has also performed simulations on different kinds of DSF, also in different fields: energy, acoustics and daylight. In the design phase of a building equipped with a ventilated double façade, it is essential to be able to predict the energy performances of the façade in the building for different design possibilities of the façade. The possibility of modelling the façade (and the building) with simulation programs can play an important role from that point of view and allows the comparison of different possible design concepts.

The prediction of the energy performances of a ventilated double façade is a complex matter. The thermal process and the airflow process interact. These processes depend on the geometric, thermo-physical, optical and aerodynamic properties of the various components of the ventilated double façade. BBRI has published a document that explains how the thermal and solar performances of VDSFs and of buildings equipped with this kind of façade can be predicted by simulation. Control aspects are considered too. In some cases, measurements and simulations have been compared (Flamant et al, 2004).

In its analysis, the objectives of the BBRI were:

- To consider not only the modelling of the VDSF alone, but also the modelling of the whole building equipped with the façade, the HVAC systems and the control aspects. Simulation programs (only software that is available in the market) are analysed. Studying the interaction between the façade, the building and the installations is important for a good assessment of the performances of VDSFs. Until now, practically no research study has assessed the impact of the control systems and the integration of VDSFs with HVAC systems;
- To analyse the capability to simulate control systems and control strategies.
- To assess the various simulation programs based on their modelling possibilities, user-friendliness, advantages, disadvantages and so on.
- To explain how a VDSF can be modelled with various software. Sometimes, 'tips' are needed. This is the reason why the knowledge of experts in simulation has been collected.

In Germany DSF is applied mostly to high-rise office buildings. The building owner or user is normally not interested in publicizing the planning information in detail. Technical journals like architectural journals often show high-quality photos of the façades and describe the usefulness of the façade with many words, but the simulation results and the measured energy consumptions or occurring temperatures are rarely presented. Additionally, detailed measurements are mostly initiated after problems occur with the indoor comfort or high energy consumptions. This leads partly to a bad reputation of DSFs among specialists in this field. Good examples of buildings with DSFs and low energy consumption as well as good indoor comfort have to be better documented, monitored in detail and publicized. Simulations need to check if the boundary conditions dependent on the user, the weather, HVAC and so on are represented in a correct way so that the reality after the erection of the building does not deviate too much from the simulated results. The experience at Fraunhofer-IBP with DSFs include the following buildings:

- Fraunhofer Central Administration, Munich (owner: Fraunhofer Gesellschaft, 2000, simulation);
- Neubau Katharinenhospital, Stuttgart (owner: city of Stuttgart, 2002, simulation);
- Neubau Bibliothek Ulm (owner: city of Ulm, 2003/2004, simulation, control strategy);
- Münchner Tor, Munich (owner: Münchner Rück, 2005, energy performance certificate);
- Sued 1, Munich (owner: Münchner Rück, 2005, energy performance certificate);
- Berlaymont Building, Brussels (owner: European Union, 2005, energy performance certificate); and
- VERU test facility for entire building concepts (owner: Fraunhofer-IBP, 2004, measurements, performance assessment calculations).

The only known simulation in Greece was made for the new headquarters building (currently under construction) of ALUMIL S. A. in Kilkis in northern Greece, as part of an international architectural competition in which the building got the second prize in the professional category. This simulation focused on the comparison of the DSF being constructed with a typical single skin façade building, a base case building following the Greek building regulations, and a typical brick building. This comparison was made on energy consumption, lighting needs in daytime, visual comfort, shading flexibility and the possibilities for views from the interior spaces of the buildings. On all accounts the proposed DSF was better than the buildings simulated, apart from the comparison on lighting needs during daytime, where the single skin building behaved better.

Measurements of the environmental performance of the existing Alumil DSF have been carried out by NKUA within the BESTFAÇADE research programme. The measurements include: dry bulb temperature (°C) of external shell, façade gap and internal shell using a thermometer; relative humidity (percentage) of external shell, façade gap and internal shell using a humidity sensor; air change rates (ach) of the façade gap, measured using a tracer gas system according to the decay method; wind speeds (m/s) externally and in the façade gap using a hot wire anemometer; global solar radiation (W/m²) perpendicular to the external shell, façade gap and internal shell, measured using pyranometers; and levels of daylighting (lux) externally, in the façade gap, internally and on task levels, evaluated by luxometer. Additionally, the energy (electricity, air conditioning, heat pumps, lighting) and environmental performance (thermal comfort, temperatures and relative humidity) of the AVAX S. A. headquarters office building have been monitored by an electronic digital system for central monitoring and control (building management system, BMS). The monitoring was carried out for the period 1 July 2000 to 30 June 2001.

Despite the interest that Portuguese architects show towards DSF technology, until recently this interest was not accompanied by the Portuguese scientific energy-related community. The situation was reversed in 2005 with the inclusion of Portuguese research institutions in scientific projects related to the evaluation of DSF technology (for example, through BESTFAÇADE). Doctorate and masters students and researchers from ISQ, LNEC and INETI (Portuguese research institutions) are currently studying different aspects of Portuguese DSF buildings, both through the use of simulation tools (Energy-Plus and DOE-2), laboratory tests (airflow field details) and gap and indoor monitoring (acoustic, thermal, lighting and energy parameters). One DSF building is currently being thoroughly monitored and more than five scientific papers where submitted for presentation at international conferences. Recently, in the context of the design of a new DSF building to be located in the Expo98 area in Lisbon, a prototype of a DSF as been build and monitored for thermal conditions in the air gap.

In Sweden simulations have been made for energy use and indoor climate of buildings with DSF, using multicell dynamic simulation tools. However, there is not yet any commercial simulation tool available that actually simulates the DSF. Knowledge is insufficient on the actual energy performance, indoor climate performance and so forth of buildings with DSFs, partly due to the fact that most of these buildings have only been in operation for a couple of years.

In many projects with DSFs simulations of temperatures, air and energy flows have been carried out before and during the design, with more or less success. Often the simulations have deviated from the result in the finished building. This reflects difficulties in defining and accurately determining the boundary conditions. To succeed with calculations not only good experience of the used simulation models is required, but also good knowledge of thermodynamics, fluid dynamics and building physics, and general shrewdness and experience of building services engineering. Increased knowledge and the improvement of simulation and calculation methods are needed.

HVAC SYSTEMS, THERMAL BEHAVIOUR, INDOOR AIR QUALITY, COMFORT AND USER ACCEPTANCE

The main heating delivery systems found in the buildings analysed for the BESTFAÇADE project were air heating followed by radiators, the main heat sources were district heating, electricity and gas or oil (Figure 1.7). For space cooling, cold air distribution

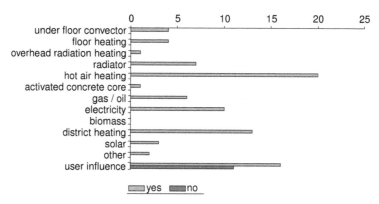

Source: Streicher (2005)

FIGURE 1.7 Types of room heating device and energy source used in BESTFAÇADE buildings

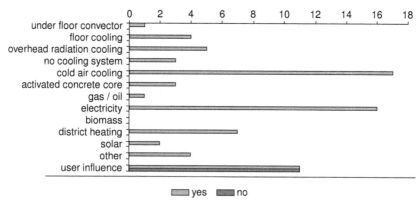

Source: Streicher (2005)

FIGURE 1.8 Types of room cooling device and energy source used in BESTFAÇADE buildings

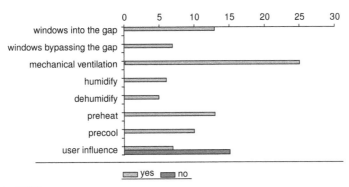

Source: Streicher (2005)

FIGURE 1.9 Ventilation and air conditioning of BESTFAÇADE buildings

dominates and is partly assisted by other appliances like cooling ceilings, floor cooling and activated concrete cooling (Figure 1.8).

Ventilation is mainly performed by mechanical ventilation, but also windows into the gap and bypassing the gap have been realized (Figure 1.9). Most of the façades have bottom and top openings in the outer shell of the façade that can be closed during winter and opened in summer (Figure 1.10). For the inner shell only half of the analysed façades have openings (mainly windows, Figure 1.11). If present, they are, of course, closable.

The airflow in the façade is mainly vertical, but also diagonal and horizontal flows have been built. Whereas most of the façades allow an airflow in summer (to avoid too high temperatures in the gap), only about half of the façades close in winter to use the air gap as unheated sun space (Figure 1.12).

The thermal comfort encountered in a building equipped with a DSF can be improved compared to a single glazed façade, especially during winter where the temperature of the inner glazing will usually be higher than a traditional façade. That reduces the thermal

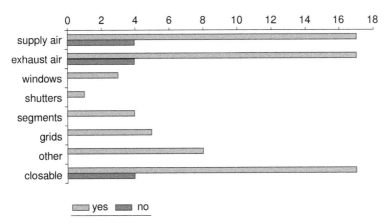

Source: Streicher (2005)

FIGURE 1.10 Ventilation openings in outer shell of analysed façades

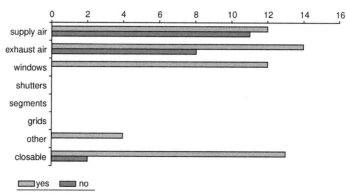

Source: Streicher (2005)

FIGURE 1.11 Ventilation openings in inner shell of analysed façades

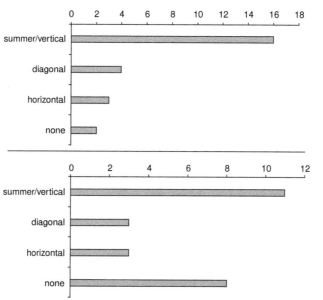

Source: Streicher (2005)

FIGURE 1.12 Airflow in the gap in summer (above) and in winter (below)

radiation of the cold surface of the glazing. In summer, the air temperature in the cavity of the DSF can be high (>50°C), depending on the type of DSF. The temperature of the inner glazing can also reach high levels (>30°C), which can create thermal discomfort and overheating (or higher energy consumption for cooling). A proper choice of the shading device and of the air ventilation rate is important.

For some types of DSF in Belgium, there is no direct influence of the façade on the air quality in the adjacent room since the air of the room is extracted via the façade (no air supply).

In some published articles on German DSF buildings, the applied technology leads to high temperatures in the façade gap in summer that partly cause overheating problems in the adjacent rooms. This is mostly solved by big air-conditioning plants and therefore high operation costs. However, some buildings show that with good planning DSFs do not necessarily lead to critical thermal situations and comfort problems. In Germany DSF planning has to be based on the summer conditions. First, the overheating problem has to be solved, and second, the façade should be adapted to possible gains during the winter. The indoor air quality may be influenced by the façade in several ways if there are openings from the rooms to the façade gap:

- positively, because in high-rise office buildings natural or hybrid ventilation might not be possible without the DSF;
- positively, if the air taken from the gap into the rooms in winter is warmer than the room temperature (possibility of reduction of the heating demand);
- negatively, as the façade may lead to bad air quality being transferred from one room to the other (for example if there are smokers); and
- negatively, if the air taken from the gap into the rooms in summer is hotter than the room temperature (increase of cooling demand).

User acceptance is dependent on these influences (thermal behaviour, indoor air quality and comfort) but also on the possibility to control the environment as well as other things such as acoustics or aesthetics.

Published data on thermal behaviour, indoor air quality and comfort of DSF buildings in Greece apply to the AVAX headquarters offices that were monitored via the building management system. Additionally, questionnaires on thermal comfort were distributed to users. The results show that:

- Due to the design, orientation and construction of the façade, good visual comfort was achieved in the office areas provided mainly by natural daylight.
- Thermal comfort was mainly described as 'neutral' with little request for changes.
- Energy consumption was reduced to almost half compared to similar buildings with conventional lighting and air-conditioning systems.

Users' acceptance of DSFs was evaluated within the BESTFAÇADE project. Currently DSF examples have no reputation in Greece because of their limited application. Initial results of the analysis show that the users are positive about DSF systems if the façade design does not lead to overheating.

Due to a lack of scientific and field studies it is difficult to report on thermal behaviour, indoor climate, comfort and user acceptance of DSF buildings in Portugal. It is also difficult to judge whether or not DSFs perform better when compared to single-glazed façades. A preliminary analysis of some of the existing buildings (type of glazing, shading) suggests that problems of overheating could occur. Information gathered from conversations with architects and maintenance personnel also point to this possibility or reality. Studies currently ongoing in Portuguese DSF buildings (within the BESTFAÇADE project, for example) will contribute to clarify these very important aspects.

The long, cold and dark winters in Sweden can cause thermal comfort problems. The low altitudes of the sun can result in fairly high cooling demand during spring and autumn. The visual comfort can be problematic due to glare in the boundary zone. For deep buildings, the daylight level can be low in the core of a building, although the façade is fully glazed.

Obviously there is an uncertainty in the building trade as to the design of buildings with highly glazed façades and how to calculate the use of energy, the comfort and the influence on these buildings of different technical solutions.

ENERGY DEMAND AND CONSUMPTION

There are very few data available for energy demand and consumption in buildings equipped with DSFs. There are publications showing very high energy consumption levels in some well-known DSF office towers in Germany. Two of the projects analysed within BESTFAÇADE include the comparison between the final energy demand calculated with DIN V 18599 according to the new Energy Performance of Buildings Directive (EPBD) requirements and the final energy consumption. In these cases both calculation of demands and monitoring of consumption by the energy provider resulted in values in the range of usual office buildings or better.

The results regarding energy consumption of DSF buildings in Greece refer to the AVAX S. A. headquarter offices. The results show that the façade design, in conjunction with the use of natural ventilation, night mechanical cooling and energy efficient lighting, results in significant energy savings and operational cost.

The level of knowledge on DSFs among most scientists, builders, developers, consulting engineers and architects in Sweden is fairly limited, especially concerning the actual energy and indoor climate performance of the building behind the façade. The exception is some major property owners/developers, engineers and architects. Portugal is also in a similar situation.

CONTROL STRATEGIES

In Belgium control systems and strategies were studied during the national project on DSFs managed by BBRI. This study showed that the control systems and strategies applied in buildings equipped with DSFs are, most of the time, very similar to those applied for single glazed buildings. An efficient operation of the façade is only possible when there is efficient control of the motorized components that are integrated in the DSF. This can be realized via the building management system (BMS), which allows an optimal operation of the different systems of the building. In Belgium the use of BMS is currently not generalized. Very often, no major differences exist in the control strategy applied between a traditional building and a building equipped with DSF. Control strategies for the façade and/or the building and plants behind the façade are variant and very dependent on the type of façade (self-operating, passive or actively influencing the climate in the building). The façade control may operate the opening of the façade (ventilation of the gap), may support the active ventilation in the gap, and may control solar shading and daylighting.

The control strategy of the building for HVAC and lighting should be adapted (or if possible linked) to the control of the façade and to the user boundary conditions. In high-rise office buildings the controls are mostly realized by the building energy management system (BEMS), which can also monitor the energy consumption of the building. Use of a BEMS makes it easier to refine the control strategy towards the most suitable and energy efficient solution and discover unnecessary energy consumption because of false control strategies or mistakes in the programming. A commissioning of the building and the plants is indispensable. In Portugal the more recent DSF buildings include façade-related control strategies, mainly for cooling and lighting systems. Control is automatic through the BEMS.

The shading system that is mostly used in the façades analysed in BESTFAÇADE is venetian blinds, with canvas screens used to a lesser extent. The control systems are nearly equally distributed among manual control, automatic control with manual override and automatic control without manual override. Daylight control is used only rarely.

INTEGRATED BUILDING TECHNOLOGY

DSFs allow, to a certain extent, the integration of technical systems for the conditioning of the rooms. Local air-conditioning systems disburden the installation ducts in the building core. With newer projects DSF developments have been realized that include, apart from the room conditioning, lighting systems and PV elements within the façade.

In Belgium, usually, for the majority of buildings equipped with DSFs, the whole concept including the façade is similar to the HVAC system. The façade is mechanically ventilated with cooling beams or cooling ceilings with activated concrete. The room air, which is extracted via the double façade, is returned via ventilation ducts to the HVAC system. The control of the shading device situated in the façade cavity can be done manually or centralized at the level of the room or at the level of the building via the BMS.

Integrated building technology exists in DSF buildings in Portugal. The oldest of these buildings, designed in the 1980s, already included a system to recover the heated gap air and use it to heat offices located far from the DSF. Figure 1.13 shows elements of building technology integrated in the façades analysed. Fire protection and active solar systems were used in about one third of the façades. Only a few buildings include PV, sound absorbers or pluvial protection devices.

COST (INVESTMENT, MAINTENANCE AND OPERATION)

In a national Belgian project on DSFs, BBRI did not carry out a very detailed analysis of the cost of the buildings equipped with DSFs. Nevertheless, the different elements having an impact on the cost of DSFs were analysed. The initial investment in the DSF bears an extra cost that can be very high for some specific types of DSF. For the most commonly used DSF in Belgium (mechanical ventilated façade with juxtaposed modules), total cost ranges from €500–700/m², including solar shading. With some types of DSF, heating appliances can be avoided or the capacity of the heating appliances can be reduced, both of which reduce the installation cost. The impact of a DSF on the dimensioning and/or the choice of the cooling systems depends on the solar performances (g-value) of the façade. The change in operation cost is proportional to energy (heating and cooling) reduction or increase for the whole building equipped with a DSF compared to a traditional building.

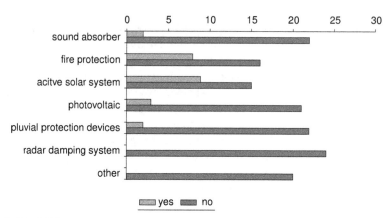

Source: Streicher (2005)

FIGURE 1.13 Integration of different devices into the façades

The maintenance cost specific to the glass skins is of course higher because of the presence of four surfaces to be cleaned. The source of the ventilation air passing through the cavity also plays a role: more cleaning is needed in the case of a cavity ventilated with outside air. The environment (pollution or no pollution) also influences the frequency of cleaning. The shading device situated in the cavity of a DSF is protected against the wind and the rain, which is favourable compared to external shading devices.

As with energy consumption, the building owner and/or users in Germany do not aim at disseminating the cost for the erection of their buildings, with or without DSFs. Construction management companies and façade manufacturers should have more insight into the investment cost. In the case of the German BESTFAÇADE project, participant Fraunhofer Institute for Building Physics is usually not party to the financial side of projects, but deals with energy efficiency and energy economy. A DSF means two façades (inner and outer shell, which does not necessarily have to add up to the price of two façades, but will lead to a higher cost than most of the usual façades with only one skin). Additionally, the DSFs are mainly glazed on both shells; glazing and especially the necessary safety glass, is more expensive than insulated panels. The investment cost of the DSF applied at the VERU test facility amounted to €1255/m² of façade area (though it should be mentioned that this façade has a very small total area of 40m²).

Figures 1.14 and 1.15 show absolute and additional costs that were collected from different publications on DSFs. Due to the wide range of technical possibilities and economic boundary conditions, a wide range of costs is reported.

For Sweden up-to-date estimated investment costs for the new WSP office building in Malmö are shown. The builder/developer is Midroc Projects, with costs according to WSP and Schüco. Approximate investment costs for different glazed façade alternatives are:

- single skin façade without exterior solar shading, €370/m²;
- single skin façade with fixed exterior solar shading (catwalk not included, simple control of solar shading included), €580/m²;

- single skin façade including daylight redirection (catwalk not included, simple control of solar shading included), €680–790/m²;
- DSF, including venetian blinds, such as the Kista Science Tower, €920–1000/m²;
- DSF box window type (cavity width 0.2m) with venetian blinds, €560/m²;
- DSF box window type (cavity width 0.2m) with venetian blinds, including daylight redirection, €610/m².

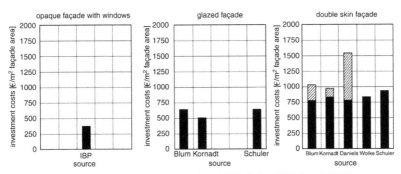

Note: The hatched areas show the range of cost mentioned in Blum (1998), Daniels (1997), Kornadt (1999), Schuler (2003) and own data.
Source: Streicher (2005)

FIGURE 1.14 Cost of DSFs compared to conventional façades

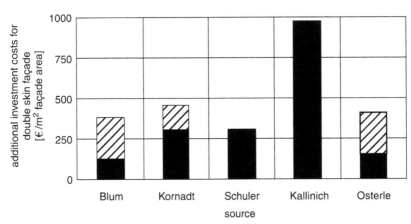

Note: The hatched areas show the range of cost mentioned in Blum (1998), Kallinich (1994), Kornadt (1999), Oesterle (2003) and Schuler(2003).
Source: Streicher (2005)

FIGURE 1.15 Additional cost of DSFs according to different authors

RESOURCE CONSERVATION AND ENVIRONMENTAL IMPACT

The environmental impact of a DSF is influenced by two factors: the additional energy needed to build the DSF (namely, in the second glass skin compared to a single glazed façade), and the reduction/increase of the energy consumption of the building. Very few data are available on this.

The environmental impact can be described in two ways: energy consumption for the operation of the building and the embodied energy used for the fabrication of the façade. Here again, two levels of façade will cause more embodied energy than one level. Besides the glazing, the DSF usually consists mainly of aluminium frames. Aluminium is a material that consumes a lot of embodied energy during the fabrication. However, the manufacturers have searched for solutions to decrease incorporated embodied energy in their product, including production of aluminium in Norway (with hybrid power) and a high recycling rate of the material.

COMPARISON TO CONVENTIONAL GLASS FAÇADES

The performance of DSFs in Austria varies intensely from buildings with good reputations, such as the Andromeda Tower, Vienna, to façades with severe problems, for example new and retrofitted buildings in Vienna where glass panes fell from the DSFs. There are, however, conventional glass façades that have poor energy performance. In some cases DSFs can be a good choice for the retrofitting of buildings constructed in the 1960s and 1970s. Advantages may be good heat storage capacity of thermal mass behind the glass façade, new aesthetics and noise reduction.

Different criteria play a role in the comparison between a DSF and a CGF. The evaluation depends on the type of DSF:

- Energy consumption for heating – few data are available. A detailed analysis must be performed in order to evaluate the possible energy savings. The DSF with juxtaposed modules is usually characterized by better thermal performances in winter than a traditional façade.
- Energy consumption for cooling – few data are available. A detailed analysis must be performed in order to evaluate the possible energy savings. For the DSF with juxtaposed modules the cooling consumption can in some cases be higher than for a traditional façade equipped with external shading devices.
- Acoustics – the acoustical insulation (against external noise) of a DSF is better. However, for specific types of DSF, problems of indirect transmission of sound through the cavity can occur (telephony effect).
- Daylight – good penetration of daylight in buildings equipped with DSFs. This is also possible with single glazed façades.
- Fire – certain types of DSF can be a problem concerning fire risk. The second skin does not allow smoke to escape.
- Shading device – a DSF allows the utilization of the shading device in all weather conditions due to the protection of the shading device situated in the façade cavity.
- Opening of windows – this is possible with certain types of DSF to allow the natural ventilation of offices, even for high buildings.
- Image – a high-tech image plays a role in the application of DSFs.

Below is a more detailed comparison of DSFs with CGFs, though it should be mentioned that DSFs may offer possibilities that cannot be realized with most conventional façades, for example natural or hybrid ventilation in high-rise buildings.

The maintenance of the façade consists of cleaning and repair. Four surfaces have to be cleaned (instead of two): the inner and outer side of the external façade, as well as inner and outer side of the internal façade. In wide DSFs (>60cm), for the two middle surfaces accessible grids are usually part of the façade gap. This facilitates the work. However, additional cleaning costs have to be taken into account with DSFs. According to the BESTFAÇADE questionnaire, the outer surface is mainly cleaned with moving platforms or cradles, whereas the inner glazing is mainly cleaned from the corridor. Also two shells might have more defects for repair compared to CGFs. By contrast, a DSF offers some advantages like a protected shading system in the gap, which will have defects less often.

The operational cost (energy cost + maintenance cost) cannot be entirely assigned to the façade system but to the building as a whole. As mentioned above, the energy consumption of a building can be negatively influenced by bad planning of the DSF, kept at the same level or also slightly positively influenced by a DSF. Accordingly the energy consumption cost will increase, stay the same or decrease.

Considering the significant number of DSF buildings in Lisbon, DSF technology is accepted among Portuguese architects and promoters. The combination of the aesthetical appearance that this technology enables and its 'environmental attributes', often mentioned in Northern and Central European specialized literature, can be one of the main reasons for its use instead of single glazed façades.

INTEGRATION OF RENEWABLE ENERGY SOURCES INTO DSFs

The only case known among the researched Austrian DSFs is the use of the concrete areas behind the south-facing glass façade as solar collectors in BiSoP, Baden, which should supply hot water radiators in the north-facing façade to reduce temperature spread. For Belgium, Greece and Portugal no applications are known.

Some DSFs include photovoltaic applications. The electrical energy generated can either be fed into the grid (in Germany with highly subsidized prices) or used in the building itself. Wind energy and solar thermal cannot easily be linked with DSFs, but of course can be an additional feature of the building. Other renewable energies like heat pumps, use of geothermal energy, wood or similar renewable fuels can be integrated into the building, partly also for preheating or precooling of the air of the building and maybe also inside the DSF gap, but they are not specifically coupled to the DSF concept.

NON-ENERGY RELATED ISSUES: ACOUSTICS, AESTHETICS, FIRE PROTECTION, MOISTURE, CORROSION, DURABILITY, MAINTENANCE AND REPAIR

These issues are often more important for user acceptance than energy-related performance because a building's energy consumption rarely affected is users directly.

Acoustics can be one of the main reasons to apply DSFs, for example with traffic noise (control room of the fire brigade in Graz, Austria, Schubertstrasse). In many cases

DSFs can reduce sound transmission from the outside due to the additional shell. However, depending on the type of DSF, problems of noise transmission from room to room by the gap can occur. This can be reduced by choosing the appropriate partitioning system or by the implementation of acoustical absorbers in the gap.

Aesthetics are often the main aspect for the application of DSFs. They give depth and a kind of 'crystal image' to the façade.

Fire protection is a serious issue with DSFs. In the case of fire, fire brigades have to destroy two shells to be able to help the building's users, also the flashover of a fire from one storey to the next can be facilitated by DSFs depending on the partitioning system. The façade manufacturers have found solutions for the second problem and, indeed, when the DSF gap is separated between each storey of the building, the problem of flashover is smaller than in conventional façades. Some types of DSFs, such as 'multi-storey DSFs', must not be applied to high buildings.

Depending on the ventilation technology, sometimes problems with condensation are reported when warm and wet exhaust air is ventilated into the gap and meets the cold inner surface of the outer glass pane (for example, FH Kufstein). Durability is also unproven due to the fact that most DSFs are prototypes, especially with pane fixtures (though the same problems may also apply to CGFs) and mechanically driven shutters or lamellae. Since DSFs are a relatively new development, there has been no scientific *in-situ* long-term analysis of a large-scale group of façades. However, problems with the durability of DSFs are unknown to date.

As mentioned above, the maintenance of the façade consists of cleaning and repair and DSFs generally incur higher cleaning costs. They may also result in higher repair costs than CGFs, but this depends on the amount and type of façade fixtures. However, because shading systems are protected they are less likely to need repair.

CONCLUSION

The BESTFAÇADE project studied 28 façades of different buildings in several partner countries by means of a standardized questionnaire. The questionnaire asked for data on location, information about the building and the façade, construction and route of airflow in the façade, as well as maintenance and cost.

Analysis of the findings took account of the aspects and types of façade in different countries, DSFs in different climatic regions of Europe, existing simulations and measurements, thermal behaviour, indoor air quality, comfort, user acceptance, energy demand and consumptions, control strategies, integrated building technology, cost (investment, maintenance and operation), resource conservation, environmental impact, comparison to CGFs, integration of renewable energy sources, and the non-energy related issues of acoustics, aesthetics, fire protection, moisture, corrosion, durability, maintenance and repair.

Most of the buildings studied were offices, though some schools and service buildings were also examined. Nearly all of the buildings had mechanical ventilation systems and both heating and cooling were performed mostly by air heating/cooling systems. The façades were mainly multi-storey and corridor types, while in Belgium juxtaposed modules are frequently used. The façade gaps were mostly naturally ventilated (except for Belgium, where the indoor air was led by mechanical ventilation via the gap to the centralized air handling unit). Shading was performed mainly with venetian blinds located in the gap. The cleaning of the outer shell was done via a cradle or a lifting platform, while the glazing of the gap was mainly cleaned from the gap or from the interior. Unfortunately, little data on energy demand and temperatures in the gap and the rooms behind the DSFs were available.

The cost of DSFs were found to be 20–80 per cent higher than CGFs and 100–150 per cent higher than opaque façades with windows. Therefore there have to be significant benefits in HVAC system costs or the operating costs of DSFs to make DSFs them more attractive than conventional façades.

AUTHOR CONTACT DETAILS

Wolfgang Streicher, Institute of Thermal Engineering (IWT), Graz University of Technology, Austria

Richard Heimrath, Institute of Thermal Engineering (IWT), Graz University of Technology, Austria

Herwig Hengsberger, Institute of Thermal Engineering (IWT), Graz University of Technology, Austria

Thomas Mach, Institute of Thermal Engineering (IWT), Graz University of Technology, Austria

Reinhard Waldner, MCE Anlagenbau Austria Gmbh & Co, Vienna, Austria

Gilles Flamant, Belgian Building Research Institute (BBRI), Belgium

Xavier Loncour, Belgian Building Research Institute (BBRI), Belgium

Gérard Guarracino, Ecole Nationale des Travaux Publics de l'Etat (LASH-DGCB), France

Hans Erhorn, Fraunhofer Institute for Building Physics (FHG-IBP), Stuttgart, Germany

Heike Erhorn-Kluttig, Fraunhofer Institute for Building Physics (FHG-IBP), Stuttgart, Germany

Mat Santamouris, University of Athens, Group of Building Environmental Studies (NKUA), Greece

Ifigenia Farou, University of Athens, Group of Building Environmental Studies (NKUA), Greece

S. Zerefos, University of Athens, Group of Building Environmental Studies (NKUA), Greece

M. Assimakopoulos, University of Athens, Group of Building Environmental Studies (NKUA), Greece

Rogério Duarte, ISQ-Instituto de Soldadura e Qualidade, Porto Salvo, Portugal

Åke Blomsterberg, University of Lund (ULUND), Energy and Building Design, SKANSKA Teknik AB, WSP Sverige AB – Sweden

Lars Sjöberg, University of Lund (ULUND), Energy and Building Design, SKANSKA Teknik AB, WSP Sverige AB – Sweden

Christer Blomquist, University of Lund (ULUND), Energy and Building Design, SKANSKA Teknik AB, WSP Sverige AB – Sweden

ACKNOWLEDGEMENTS

With the support of Intelligent Energy Europe (EIE/04/135/S07.38652).

The sole responsibility for the content of this report lies with the authors. It does not represent the opinion of the European Communities. The European Commission is not responsible for any use that may be made of the information contained therein.

REFERENCES

BBRI (2004) 'Ventilated double façades: Classification and illustration of façade concepts', Belgian Building Research Institute, Department of Building Physics, Indoor Climate and Building Services, Brussels, Belgium

Blomsterberg, Å. (2003) 'Project description, glazed office buildings: Energy and indoor climate', Lund University, Sweden, www.ebd.lth.se

Blum, H. J. (1998) 'Das innovative Raumklimakonzept', *Bauphysik*, vol 20, Heft 3, pp81–86

Carlsson, P.-O. (2003) *Glazed Façades – Double Skin Façades*, Arkus, Stockholm (in Swedish)

Crespo, A. M. L (1999) *3x2: Three Takes on Double Skins*, Harvard University, Cambridge, MA

Daniels, K. (1997) *The Technology of Ecological Building: Basic Principles and Measures, Examples and Ideas*, Birkhäuser Verlag, Basel

Flamant, G., Heijmans, N. and Guiot, E. (2004) 'Ventilated double façades: Determination of the energy performances of ventilated double façades by the use of simulation integrating the control aspects – Modelling aspects and assessment of the applicability of several simulation software', Belgian Building Research Institute, Department Building Physics, Indoor Climate and Building Services, Brussels, Belgium

Harrison, K. and Meyer-Boake, T. (2003) *The Tectonics of the Environmental Skin*, University of Waterloo, School of Architecture, Ontario, Canada

Kallinich, D. (1994) 'Doppelfassaden', *Beratende Ingenieure*, vol 9, September, pp36–45

Kornadt, O. (1999) 'Doppelfassaden: Nutzen und Kosten', *Bauphysik*, vol 21, pp10–19

Le Corbusier, C. E. J. (1964) *Feststellungen zu Architektur und Städtebau*, Ullstein, Berlin, Frankfurt and Vienna

Neubert, S.(1999) *Doppelfassaden – Ein Beitrag zur Nachhaltigkeit?*, Ecole Polytechnique Federale de Lausanne, Lausanne

Oesterle, E. (2003) 'Doppelfassaden - Möglichkeiten und Grenzen der zweiten Gebäudehülle', *CCI Fachtagung 'Doppelfassaden'*, Karlsruhe, 27 November

Poirazis, H. (2004) 'Double Skin Façades for Office Buildings – Literature Review', Division of Energy and Building Design, Department of Construction and Architecture, Lund Institute of Technology, Report EBD-R–04/3, 2004, Lund University, Lund

Saelens, D. (2002) 'Energy performance assessment of single storey multiple-skin façades', dissertation, Katholieke Universiteit Leuven, Faculteit Toegepaste Wetenschappen, Arenbergkasteel, B-3001, Catholic University of Leuven, Leuven

Schuler, M. (2003) 'Ausgeführte Doppelfassaden mit minimierter Gebäudetechnik', CCI Fachtagung 'Doppelfassaden', Karlsruhe, 27 November 2003

Streicher, W. (ed) (2005) WP 1 Report 'State of the Art', BESTFAÇADE, Best Practice for Double Skin Façades, EIE/04/135/S07.38652

Svensson, A. and Åqvist, P. (2001) 'Double skin glazed façades: Image or a step towards a sustainable society?', Arkus, Stockholm (in Swedish)

2

Artificial Intelligence in Buildings: A Review of the Application of Fuzzy Logic

D. Kolokotsa

Abstract
A review of fuzzy logic applications to buildings research is the topic of this paper. Emphasis is given to the applications that deal with the regulation and modelling of indoor thermal comfort, visual comfort and indoor air quality. The improvement of indoor comfort with simultaneous energy conservation is considered. Heating, ventilation and air-conditioning (HVAC) systems operation, fault diagnosis and their modelling by fuzzy logic for prediction of their behaviour is also investigated. Significant attention is provided to the regulation of the indoor environment by taking into account the combining effect of all comfort aspects.

■ *Keywords* – energy management; indoor environment; fuzzy controller; thermal comfort; visual comfort; indoor air quality

INTRODUCTION

Artificial intelligence (AI) is defined as intelligence exhibited by an artificial entity. AI generally assumes the presence of computers. Neural networks, genetic programming, fuzzy logic, computer vision, heuristic search, Bayes networks, planning, language understanding and combinations any of the above are AI's technologies (Nilsson, 1998).

AI in buildings technology has been investigated during recent decades. Modern control systems provide optimized operation of the energy systems while satisfying indoor comfort. Recent technological developments based on AI techniques offer several advantages compared with the classical control systems. The use of artificial neural networks (ANNs) in various applications related to energy management has been growing significantly over the years (Bellas-Vellidis et al, 1998). Current applications are related to energy demand forecasting and to HVAC systems of buildings (Curtiss et al, 1993; Kreider, 1995; Huang and Lam, 1997; Khotazad et al, 1997; Han et al, 1997; Karatasou et al, 2006). The results have revealed the potential usefulness of the neural networks for the energy management of houses and buildings. Evolutionary computing, namely, genetic algorithms (GA) are employed in buildings since they have proved to be robust and efficient in finding near-optimal solutions in complex problem spaces (Wright et al, 2002; Kolokotsa et al, 2003).

Fuzzy logic (Figure 2.1) was introduced by Zadeh (1973) as a mathematical way to represent ambiguity and vagueness. Zadeh's approach has gained significant attention since then (Dubois and Prade, 1980). For the last two decades fuzzy logic has been extensively applied to buildings, targeting the improvement of the indoor comfort conditions, detailed modelling of thermal comfort indices, reduction of energy consumption and adjustment, and optimization of HVAC systems operation (Dexter and Trewhella, 1990; Dounis et al, 1993; Angelov, 1999; Angelov et al, 2000a; 2000b; 2003; 2005).

The application of artificial intelligence in buildings has been reviewed by researchers in the past, where background information on AI methods and review of knowledge-based expert systems in various aspects of buildings research are presented (Loveday and Virk, 1992; Maor and Reddy, 2003; Krarti, 2003). Moreover, neural networks applications for buildings energy research are reviewed (Kalogirou, 2000; Bailey and Curtiss, 2001).

The aim of this paper is to investigate and review the applications of fuzzy logic on buildings technology in terms of improvement of indoor comfort and energy conservation.

Indoor environment is assessed by the following parameters:

- Thermal comfort – indoor comfort control is one of the most important tasks of intelligent buildings. Among indoor climate characteristics, thermal comfort is of major importance. Thermal comfort is influenced by a great number of parameters such as temperature, humidity, indoor air velocity, radiant temperature, metabolic rate and the insulation of the clothing (Fanger, 1970; Olesen and Parsons, 2002).
- Indoor air quality is also of major importance in modern buildings. It is controlled either at the design stage by reducing possible pollutants in the room or during operation by the use of ventilation system. Major indoor pollutants are carbon dioxide (CO_2) concentration and volatile organic compounds (VOCs) (Allard et al, 1998; ASHRAE, 1999).
- Indoor visual comfort is another major concern in intelligent buildings influenced by illumination, brightness, contrast, glare and psycho-physiological sensations, for example the quantity, distribution and quality of light (CIBSE, 2004).

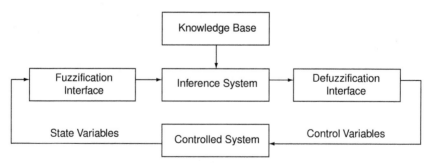

Source: Alcala et al (2005)

FIGURE 2.1 The structure of the fuzzy logic controller

- Energy consumption.
- HVAC system status and stable operation.
- Outdoor climate parameters.

All the above parameters are interlinked. For example indoor thermal comfort is influenced by the ventilation requirements and all influence energy consumption and cost. The first attempts of introducing fuzzy logic to buildings intelligence in the early 1990s focused on just one of the above parameters. Recent work deals with the application of fuzzy logic to more than one field (that is, the combination of thermal comfort and indoor air quality). Moreover, other AI techniques are also used in combination with fuzzy logic to provide more innovative applications and results. The most commonly used AI techniques are genetic algorithms and neural networks.

In the next sections the applications and results of fuzzy logic in the improvement of comfort are analysed. The analysis is divided into three main categories: fuzzy logic and comfort, fuzzy logic energy and building services, and fuzzy logic and hybrid applications.

FUZZY LOGIC AND INDOOR THERMAL COMFORT

There are two main research areas to which fuzzy logic has been applied to the topic of thermal comfort: modelling of the thermal sensation and control of the thermal environment. These research areas are analysed below.

MODELLING OF THE THERMAL SENSATION
State of the art

Thermal sensation is influenced by a great number of parameters that are either objective or subjective (Fanger, 1970; Olesen and Parsons, 2002). Human thermal comfort is very difficult to model mathematically because the real process in human perception is still unknown and far too complex to comprehend. Fuzzy logic techniques are dedicated to the expression of the ambiguity in human thinking, subjectivity and knowledge in a comparatively accurate manner. Therefore the use of fuzzy logic in thermal comfort's representation is of major importance.

Fuzzy logic in thermal sensation is used by the following:

- Feriadi and Hien (2003) used it for modelling the thermal environment of tropical naturally ventilated residential buildings. A survey was performed and the predicted comfort votes using fuzzy logic were close to the survey votes of building occupants.
- Chen et al (2006a) used possibilistic fuzzy regression and least squares fuzzy regression to model thermal comfort in air-conditioned rooms.
- Fuzzy adaptive networks were used by Chen et al (2006b) to model thermal comfort. A neuro-fuzzy approach was selected to represent the vagueness of thermal comfort.
- Shimizu and Jindo (1995) employed fuzzy regression to analyse the sensitivity data for a vehicle interior. They studied the influence of airflow velocity, temperature and sun load on the thermal sensation. Optimum membership functions were obtained that showed the non-linear relationship between each physical variable (airflow velocity, temperature and sun load) and thermal sensation.

- Wen and Zhao (1998) predicted human psychological response in a dynamic thermal environment based on fuzzy set theory. Numerical results were compared with the available experimental results, verifying the effectiveness of the algorithm.
- Hamdi et al (1999) estimated the thermal comfort level based on fuzzy logic. The thermal comfort level was considered to depend on air temperature, mean radiant temperature, relative humidity, air velocity, activity level of the occupants and their clothing insulation. The fuzzy thermal sensation index was calculated by linguistic rules that describe human comfort level as the result of the interaction of environmental variables with the occupant's personal parameters. The fuzzy comfort model was deduced on the basis of Fanger's 'predicted mean vote' (PMV) equation (Fanger, 1970).
- Ari et al (2005) focused on individual thermal satisfaction with the development of a new thermal comfort measure: the 'degree of individual dissatisfaction' (DID), based on Fanger's PMV and percentage of people dissatisfied (PPD) (Fanger, 1970). In this work, a gradient-based scheme was used to minimize energy consumption by varying office temperatures while keeping the overall population dissatisfaction to less than 10 per cent. Fuzzy rules were generated by data using a gradient-based optimization. The fuzzy logic control scheme based on nearest neighbours approximated closely the gradient-based optimized results.

Table 2.1 is a summary of the state of the art.

Discussion

The subjective feelings or perception of thermal comfort cannot be mathematically defined and thus are vague. Many ergonomic systems have this humanistic nature; they can be defined as systems whose behaviour is strongly influenced by human judgements, perceptions or emotions. Fuzzy approaches are ideally suited to handle and model systems with strong humanistic natures. This is explained by the closeness of fuzzy logic surveys with human surveys. Fuzzy logic is suitable for general modelling of thermal comfort but additional techniques should be applied in order to 'fit' the general thermal fuzzy model to each occupant's perception.

TABLE 2.1 Summary of fuzzy logic in thermal sensation modelling

AI TECHNIQUES	PERFORMANCE AND REMARKS	REFERENCES
Fuzzy modelling	The difference between Fanger's PMV (Fanger, 1970) and the fuzzy PMV values is very small	Feriadi and Hien, 2003; Hamdi et al, 1999; Wen and Zhao, 1998
Neuro-fuzzy adaptive networks	The neuro-fuzzy approach represents linguistically vague issues regarding comfort	Chen et al, 2006b
Fuzzy regression analysis	Extraction of optimum membership functions. The non-linearity is revealed	Chen et al, 2006a; Shimizu and Jindo, 1995
Fuzzy gradient-based optimization	Minimization of individual people dissatisfied	Ari et al, 2005

CONTROL OF THE THERMAL ENVIRONMENT
State of the art
The second major research area for fuzzy logic in thermal comfort is the control of the thermal environment. The following research is dedicated solely to thermal comfort control:

- Dounis et al (1992; 1995c) and Dounis and Manalokis (2001) proposed various designs for thermal comfort controllers based on PMV index's regulation. These controllers can be used in any building without the necessity of a building model.
- Moreover, Dounis et al (1995a) analysed the performance of the fuzzy controllers for thermal comfort regulation versus classical controllers – on-off, proportional integral derivative (PID) and so on. The same comparison was performed by Gouda et al (2000).
- Yonezawa et al (2000) introduced a fuzzy rule base to maintain room temperature under the PMV comfort basis.
- Piao et al (1998) used a simple fuzzy adaptive controller to achieve temperature control in HVAC.
- Chu et al (2005) proposed a least enthalpy estimator (LEE) that combines the definition of thermal comfort level (TCL) for load prediction in order to provide timely suitable settings for a fan coil unit (FCU) fuzzy controller used in HVAC. The controller is divided into two fuzzy rule sets where the inputs of the first fuzzy rule set express room thermal load. The second fuzzy rule set compensates the effects of outdoor weather disturbance and indoor heat gain. The proposed algorithm attains 35 per cent reduction of energy consumption compared to a room thermostat.
- A fuzzy PID controller for the control of the PMV index was introduced by Calvino et al (2004). No models of the indoor or outdoor environment are required. Instability problems are pinpointed and avoided.
- Egilegor et al (1997) developed a neuro fuzzy control system that regulates the fan coils airflow rate of three zones of a dwelling to improve comfort using a fuzzy proportional integral (PI). The controlled variable is used in the PMV index. The neural network uses as the adaptive parameter an offset of the PMV error and learns to associate a change in the offset to a change of the PMV index mean value. The PMV oscillations are quite strong while the output of the fuzzy controller is an analogue signal, the fan coils have three discrete flow rates. No results regarding the energy consumption are presented.
- A fuzzy controller for the regulation of indoor temperature of a discontinuously occupied building is compared with a classical controller by Fraisse et al (1997b). The fuzzy controller combines the variables of thermal comfort (set point temperature of the occupied space and set point temperature of the unoccupied space), the programming of occupancy scenarios and the optimization of the restart time of the heating system, to regulate the indoor temperature by controlling the supply temperature of the water in the heating system.

- A quasi adaptive fuzzy controller for space heating in passive solar buildings that is responsive to the lagging effects of solar energy inputs is proposed by Gouda et al (2006). The controller is divided into two main modules: a conventional static fuzzy controller and feed-forward neural network with a singular value decomposition (SVD) algorithm. An estimation of the internal air temperature at least one hour ahead in time and to within typical measurement uncertainty is provided by an ANN. Experimental results of the fuzzy controller are compared to simulations of the conventional PI heating system. The fuzzy controller follows the variable set point more accurately and reduces the afternoon overheating significantly compared to the conventional control problems.
- Moreover, Gouda et al (2001) used the PMV to control the indoor temperature of a space by setting it at a point where the PMV index becomes zero and the predicted PPD achieves a maximum threshold of 5 per cent compared to a tuned PID controller.

Table 2.2 presents a summary of the analysis.

Discussion

Almost all researchers evaluate the performance of the fuzzy controllers for the regulation of the indoor thermal comfort versus classical controllers. Another criterion for the evaluation of thermal comfort control using fuzzy logic is the energy conservation achieved. In all cases fuzzy control reduces oscillations and overshootings, resulting in

TABLE 2.2 Summary of fuzzy logic control in thermal comfort applications

AI TECHNIQUES	PERFORMANCE AND REMARKS	REFERENCES
Fuzzy controller	Temperature and PMV regulation is achieved within the comfort levels. The fuzzy controller predicts accurately the weekend and night set backs. Experimental and simulation analysis	Dounis et al, 1992; 1995c; Fraisse et al, 1997b; Yonezawa et al, 2000; Dounis and Manalokis, 2001; Gouda et al, 2001; Calvino et al, 2004
Fuzzy controller for thermal load prediction	99.84% of thermal comfort while 58.75% is achieved with a thermostat. There is a 35% reduction of energy consumption compared to a room thermostat. Experimental analysis is performed	Chu et al, 2005
Fuzzy adaptive controller	Better tracking and overheating reduction compared to PI heating controller	Piao et al, 1998; Gouda et al, 2006
Neuro-fuzzy controller	Strong PMV oscillations. No results regarding the energy consumption are presented	Egilegor et al, 1997

optimum control of thermal variables and minimizing energy consumption. The reduction of energy use can be more than 15 per cent when compared to conventional control methods.

FUZZY LOGIC AND INDOOR AIR QUALITY

Fuzzy logic has a significant number of applications for ambient air quality and air pollution management and control. Indoor air quality is mainly tackled in combination with the other parameters of indoor comfort, that is, indoor thermal comfort and visual comfort. The research work that investigates hybrid techniques for indoor comfort, building services and energy systems is analysed in below.

The research work dedicated solely to indoor air quality is as follows:

- Dounis et al (1996b) analysed the performance of a fuzzy controller for the regulation of indoor air quality in naturally ventilated buildings. The analysis was based on simulations. CO_2 concentration, which was the controlled variable, was kept at satisfactory levels, while a good stability of the control parameter (window opening) was achieved.
- A device targeted to the quantification of carbon monoxide and nitrogen dioxide in mixtures with relative humidity and volatile organic compounds by using an optimized gas sensor array was proposed by Zampolli et al (2004). In this work, a fuzzy logic system was used for pattern recognition to identify and discriminate concentrations as low as 20 parts per billion for NO_2 and 5 parts per million for CO.
- Various techniques for the control of the indoor air quality of naturally ventilated buildings are compared by Dounis et al (1996a) including on-off, PID, PI with deadband and fuzzy control. The fuzzy control technique reduces the oscillations of the controlled variable and simulation results prove the technique to be the most promising for the indoor air quality control of naturally ventilated buildings.

FUZZY LOGIC AND VISUAL COMFORT
STATE OF THE ART

Visual comfort involves mainly daylighting and electric lighting control through the modulation of shading devices, dimmers and so on. Lighting control has now become an essential element of good design and an integral part of energy management programmes.

The research work focusing on visual comfort and lighting control is the following:

- Dounis et al (1993; 1995b) proposed a fuzzy control scheme for visual comfort in a building zone. In the proposed controller the indoor illuminance and the daylight glare index are regulated through the adjustment of shading and electric lighting. The response of both the daylight glare index and indoor illuminance are quite satisfactory.
- Lah et al (2006) dynamically regulated the illuminance response of built environment in real-time conditions targeting to assure the desired inside illuminance with smooth roller blind moving. A fuzzy Takagi-Sugeno type controller was developed and tested

in real-time operation. The illuminance fuzzy controller gives the best control performance, assures the inside daylight illuminance with moderate continuous roller blind movement while the desired value deviates up to ±20 lux.

- Guillemin and Molteni (2002) developed an event-oriented fuzzy controller to integrate users' wishes with lighting control. The adaptation is performed using GA and has two goals: to learn wishes of new users concerning blind position, and to accumulate experience concerning previous learned wishes and energy efficient control. The system is energy efficient based on the original fuzzy rules and a filter prevents learning of bad energy wishes, as long as these wishes are not repeated by the user.
- Kurian et al (2005) focused on the adaptive neuro-fuzzy predictive control of artificial light in accordance with the variation of daylight. Simulated data by the software package RADIANCE are incorporated into a computational model. The simulations provide the knowledge by which a supervised learner, implemented in an adaptive neuro-fuzzy inference system, is trained for faster predictions.
- The objective of Sulaiman et al (2005) was to apply fuzzy logic control to lighting and dimming of passive fibre optic daylighting systems. Large diameter solid core optical fibres are used to collect, concentrate and distribute daylight passively throughout residential or commercial buildings. This arrangement contributes to the reduction of the energy consumption required for lighting the interiors of building structures. The fuzzy logic control, shown to be successful in maintaining the illuminance of a suitably sized model room of a house, proves that energy power savings and energy efficiency can be achieved using the proposed technology.

Discussion

Control of lighting and visual comfort preservation is not as straightforward a task as it first appears. Users usually prefer daylighting unless they experience glare problems that are very difficult to define by the presence of sensors. Moreover, daylighting is preferable for energy efficiency reasons. Fuzzy control is mainly applied in order to provide sufficient illuminance to fulfil users' requirements while minimizing the use of electric lighting. As is mentioned below, the reduction of electric lighting energy use can contribute to more than 50 per cent of energy conservation, when compared with no control.

FUZZY LOGIC ENERGY CONSUMPTION AND HVAC SYSTEMS
STATE OF THE ART

An HVAC system (Figure 2.2) is essential to a building in order to keep its occupants comfortable. The difficulties of obtaining good control over HVAC systems that possess multivariables, long delay times and non-linear characteristics by using traditional control methods are well known (Levermore, 2000; Kreider et al, 2002). Moreover, many HVAC systems do not maintain a uniform temperature throughout a building's zone because of the lack of intelligent systems.

Fuzzy control is an efficient way to implement the control process in HVAC and energy systems and quite significant efforts have been carried out for that purpose. The research work is divided into three main categories: energy planning and load prediction, control of building services, and fault detection and diagnosis of building services.

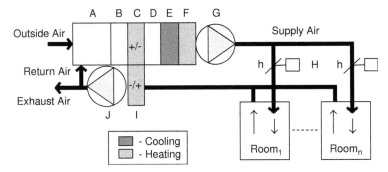

Notes: A – This module mixes the return and the outside air to provide supply air, and also closes outside air damper and opens return air damper when fan stops. B – This is a filter to reduce the outside air emissions to supply air. C – The preheater/heat recovery unit preheats the supply air and recovers energy from the exhaust air. D – A humidifier raising the relative humidity in winter. E – This is a cooler to reduce the supply air temperature and/or humidity. F – An after-heater unit to raise the supply air temperature after humidifier or to raise the supply air temperature after latent cooling (dehumidifier). G – The supply air fan. H – The dampers to demand controlled supply airflow to rooms. I – This is a heat recovery unit for energy recovery from exhaust air. J – The exhaust air fan.
Source: Alcala et al (2005)
FIGURE 2.2 The structure of an office building HVAC

For energy planning and energy load prediction:

● Kubota et al (2000) developed a prediction system based on genetic programming and fuzzy inference systems. Genetic programming is applied for the feature extraction and selection, and fuzzy inference is used for the building energy load prediction. The method is compared to the Kalman filtering algorithm and a feed-forward neural network (NN) with four layers. Although the NN is better for load prediction, the proposed method can extract meaningful information from the measured data and can predict the building energy load of the next day.
● Michalik et al (1997) modelled energy-using behaviour of residential customers by the use of a fuzzy logic approach. Fuzzy filters are applied to transfer uncertainties of customer declarations expressed in linguistic variables into parameters of structural models. The application of fuzzy filters provides average patterns of energy consumption with time averaging. The fuzzy filters differ significantly from the statistical approach where averaging is carried out across a sample of customers. This method allows reduction of sample sizes in surveys, reducing costs of model development.
● A fuzzy multi-objective energy resource allocation programme was formulated by Jana and Chattopadhyay (2004) with the following three objectives: minimization of total cost of direct energy, minimization of use of non-local sources of energy, and maximization of overall efficiency; that is, the minimization of total energy use of domestic lighting in order to serve rural planning objectives.
● A fuzzy model of building thermal analysis was developed by Skrjanc et al (2001). The fuzzy model is based on experimental data and is based on the Sugeno structure. It has two inputs (global solar radiation and outdoor temperature) and one output

(indoor temperature). The fuzzy modelling is compared to theoretical modelling based on energy balances expressed by differential equations. The fuzzy model is superior as it is not complicated and does not include time-consuming procedures. Moreover, no simplifications are necessary.

For the control of building services:

- Dexter and Trewhella (1990) created an introduction to the use of fuzzy sets in a building services application and illustrated the subjective nature of the evaluation process and the difficulties of assessing the overall performance of a building control system in a realistic manner. They also demonstrated how the tuning of the controller affects the different aspects of its overall performance.
- Zhang et al (2003a) proposed a new method for fuzzy logic reasoning in HVAC systems: the functioning fuzzy-subset inference (FFSI). They compared the FFSI method with the compositional rule of inference (CRI) method that is widely used, and with a traditional PID controller. Moreover, they developed a testing room dynamic thermal system (TRDTS) model in order to check the requirements and the performance of the FFSI in HVAC and they analysed the dynamic responding characteristics of the model in variable water volume systems (Zhang et al, 2003b). A stable error that occurs in CRI is not present in FFSI. FFSI is more sensitive and experiences more oscillations. FFSI performs better than PID. The FFSI control of the indoor temperature has rapid responding speed, high control precision, good stability and a small, stable error.
- A fuzzy module was developed by Reyes-Garcia and Corona (2003) for the control of ventilation and air conditioning. The module was evaluated by experts and the evaluation results were satisfactory. Moreover, HVAC fuzzy principles are given by Lea et al (1996). The performance of the proposed control system from a subjective point of view of people living in the facility was evaluated.
- A Sugeno type fuzzy model for a FCU was developed and tested by Ghiaus (2001). The fuzzy model was compared to a linear model versus the initial system and was found to be much closer to the initial system than the linear model. Moreover, similarity analysis was applied to fuzzy models of FCUs for HVAC systems by Singh et al (2005) in order to reduce complexity. These fuzzy models allow qualitative interpretation and faster computations, without deteriorating the prediction accuracy.
- A fuzzy logic controller (FLC) was developed by Chao et al (2000) to satisfy a broad spectrum of installation sizes of heating, cooling and ventilation without any modification. Simulations were conducted using the same FLC in a greenhouse and a broiler house. Disturbances investigated included variations in outside temperature, internal heat load (or solar load), and step changes in set point temperature. FLC system responses were compared with a conventional controller for stability, overshoot and mean square error from set point temperature. The FLC provided improvements in performance and reduced the mean square error.
- A hybrid method fuzzy logic and PID controller was developed by Rahmati et al (2003). Simulation results show that this control strategy is robust. The response of the hybrid controller was compared with a classical PID controller.

- A multiple model predictive control (MMPC) strategy based on Takagi Sugeno fuzzy models for temperature control of the air handling unit (AHU) in HVAC systems is presented by He et al (2005). The AHU system is divided into a set of Takagi Sugeno models. Better performance is achieved through the MMPC comparing to the PID controller and oscillations were diminished.
- So et al (1997) applied a self-learning fuzzy controller to an AHU. While PID control for AHU is a simple and straightforward solution, fuzzy control can be more robust, more energy efficient and faster in responding to changes due to executing expert knowledge. Additionally, the self-learning fuzzy controller has the advantage of being adaptable to changes in the control process and the environment, and no system model needs to be assumed beforehand. Therefore faster response and reduction of the energy consumption of the AHU is achieved.
- A self-tuning fuzzy controller was tested in the heating mode by regulating the steam flow rate to the steam coil by Huang and Nelson (1999). The refinement of the initial fuzzy rule set is necessary for optimum performance (Huang and Nelson, 1994a; 1994b). The initial rule set was represented by a look-up table and updated based on performance measures that correspond to overshoot, rising time, steady-state error and so on. The overshoot is eliminated and the rise time only increased slightly. Therefore, it is possible to achieve excellent dynamic process control with less effort by using a self-tuning fuzzy controller.
- A self-regulating fuzzy algorithm was employed to control the air-conditioning system by Zheng and Xu (2004). Qualitative and quantitative variables, system error and error change are used for weights in the fuzzy algorithm. The self-regulating fuzzy controller, compared with a non-regulated fuzzy controller, has shorter response times, less stable error and better quality of self-adaptation to the changes of the system's parameters.
- A Takagi Sugeno fuzzy model was developed for identification purposes and prediction for temperature control in an air-conditioning system by Sousa et al (1997). The fuzzy model acts as a predictor to compensate for disturbances and modelling errors. Comparisons with a non-linear predictive control scheme show that the fuzzy model requires fewer computations and achieves better performance.
- A neural-fuzzy based sensor is presented by Liao (2005) to be used for close-loop boiler control schemes. Both simulation and experimental results show that the proposed technique results in almost a 12 per cent energy saving and improvement in the control of thermal comfort in the built environment, when compared to conventional boilers.
- Liu and Dexter (2001) developed a fault-tolerant supervisory control (FTC) model that uses fuzzy models to predict the control performance. The performance of the fault-tolerant control was analysed based on the energy consumption when the set points are constant and when they are selected by the FTC. The FTC improves the energy efficiency compared to constant set point temperature.
- Moreover, Fraisse et al (1997a; 1999) studied a significant number of optimization methods of heating restart time in intermittently occupied buildings. The duration of

the recovery period was calculated either directly from internal and/or external temperatures or indirectly and repeatedly by means of a building model. The various methods studied are based on classical analytical functions, with the exception of one method that is based on fuzzy logic. The energy consumption and the PPD corresponding to fuzzy logic is lower than all other methods.

- A neuro-fuzzy optimization method for the in-building section of centralized HVAC systems was also proposed by Lu et al (2005). The duct and pipe network of the HVAC system was modelled using adaptive neural fuzzy inference systems (ANFISs) and optimal differential pressure (DP) set points based on limited sensor information were extracted. Genetic algorithms are applied for the minimization of the total power consumption of the section. The method is compared with other methods where the temperature of the chilled water supply is fixed, while the temperature of the chilled water supply in the proposed method is optimal.

Research on fault detection and diagnosis in HVAC systems:

- Ngo and Dexter (1999) described a model-based approach to diagnosing faults in AHUs that avoids false alarms caused by sensor bias but do not require application-dependent thresholds to be selected. Test data are used to identify a partial fuzzy model that describes the steady-state behaviour of the equipment at a particular operating point. The partial model is then compared to each of the reference models using a fuzzy matching scheme to determine the degree of similarity of the partial model and the reference models. The fuzzy reference models describe in qualitative terms the steady-state behaviour of a particular class of equipment with no faults present and when each of the faults has occurred.
- A method of diagnosis has been proposed by Dexter and Benouarets (1997) that uses fuzzy models and fuzzy matching to account for modelling errors and the ambiguity that arises when different faults, or faulty and fault-free operation, exhibit similar symptoms. The use of generic fuzzy reference models was also suggested by Dexter and Benouarets to reduce the false alarm rate in information-poor systems, though the results of the diagnosis are inevitably more ambiguous.
- The presence of faults was detected by a fuzzy model-based fault diagnosis scheme that uses generic reference models to describe the behaviour of the plant when it is operating correctly and when one of a predefined set of faults has occurred (Ngo and Dexter, 1998). Experimental results are presented that demonstrate the use of the tool to remotely commission the cooling coil of an AHU in a commercial office building.
- Kolokotsa et al (2006) proposed a methodology for detecting sensor failures in building energy management systems. The fault diagnosis decision criterion is the average absolute prediction error between the actual and the predicted values of the sensor. The predicted value was calculated by a model based on faultless operation data collected using fuzzy control. Three experiments are presented with simulated biases in the temperature, illuminance and CO_2 sensors. The fault detection is quite satisfactory with minimum false alarms.

Table 2.3 summarizes the research outlined above.

TABLE 2.3 Summary of fuzzy logic in energy consumption and HVAC systems

APPLICATION	AI TECHNIQUES	PERFORMANCE AND REMARKS	REFERENCES
Energy planning and energy load prediction	● Fuzzy genetic programming ● Fuzzy filtering ● Fuzzy multi-objective optimization of energy planning ● Fuzzy model	Accurate prediction of the energy load one day ahead. The fuzzy model of the thermal analysis is superior to the model based on differential equations	Michalik et al, 1997; Jana et al, 2004; Kubota et al, 2000; Skrjanc et al, 2001
Fuzzy control of HVAC systems	● Fuzzy controller ● Fuzzy-subset inference method ● Self-learning fuzzy controller ● Self-tuning fuzzy controller	Better responding speed, shorter response time, good stability and small steady-state error compared to conventional and PID controllers. The self-learning and self-tuning fuzzy controllers have better performance than the initial fuzzy controllers. Also better quality of self-adaptation is achieved through self-tuning and self-learning	Dexter and Trewhella, 1990; Huang and Nelson, 1994a; 1994b; 1999; Lea et al, 1996; So et al, 1997; Chao et al, 2000; Liu and Dexter, 2001; Rahmati et al, 2003; Reyes-Garcia and Corona, 2003; Zhang et al, 2003a; 2003b; Zheng and Xu, 2004; He et al, 2005; Liao, 2005;
Fuzzy model of HVAC systems	● Fuzzy model of temperature control of AHU ● Fuzzy model for fan coil unit ● Fuzzy model for prediction of the control performance	The fuzzy models are mainly used as predictors in order to reduce disturbances. Fuzzy models are closer to the real measurements than linear models of HVAC systems. The fuzzy model can reduce the tuning procedure	Dexter and Benouarets, 1997; Ngo and Dexter, 1998; Ghiaus, 2001; Liu and Dexter, 2001; Reyes-Garcia and Corona, 2003; Zhang et al, 2003b; He et al, 2005; Kolokotsa et al, 2006
Optimization	● Neuro-fuzzy optimization method	10% energy conservation compared to non-optimized systems (i.e. fixed temperature of the chilled water supply, etc)	Fraisse et al, 1997a; 1999; Lu et al, 2005
Fuzzy fault detection of HVAC systems	● Faults of AHU ● Fuzzy models for fault detection	Fuzzy models are used to describe the operation of the plants under normal conditions and provide normal operation data	Dexter and Benouarets, 1997; Ngo and Dexter, 1998; 1999; Kolokotsa et al, 2006

Discussion

Fuzzy logic is extensively used in either control or modelling of building services or energy consumption of buildings. The control of HVAC systems is superior to conventional control strategies (namely, on-off and PID control) according to the above analysis, in almost all performance criteria. Moreover, self-adapting fuzzy control or self-tuning fuzzy control leads to more robust results than the conventional and initial (before adaptation and tuning) fuzzy sets.

HYBRID APPLICATIONS

As mentioned in the introductory section, indoor comfort and energy consumption are completely coupled. This section outlines the research work that incorporates more than one aspects of comfort and energy conservation:

- Alcala et al (2005) focused on thermal comfort and indoor air quality regulation while minimizing energy consumption. In their work the contribution of GAs for fuzzy logic controllers and systems stability is also investigated. GAs are introduced for rules selection and weights adjustment targeting to improve the controller's accuracy by selecting the set of rules best to obtain simpler, and thus easily understandable, FLC by removing unnecessary rules. They demonstrated a 10 per cent improvement of the overall performance (energy and stability) compared to on-off controllers.
- Fuzzy control strategies and supervisory techniques for temperature, humidity and CO_2 concentration were studied by Shepherd and Batty (2003). Their work is compared to conventional PID controllers. A 15 per cent improvement of energy consumption compared to PID controllers and a reduction of the steady-state error for temperature and humidity were demonstrated.
- Eftekhari and Marjanovic (2003) developed a fuzzy controller for naturally ventilated buildings to provide thermal comfort and adequate air distribution inside a single-sided naturally ventilated room. The inputs to the controller are the outside wind velocity and direction, and outside and inside temperatures. The output is the position of the opening. The controller is evaluated in response to the outdoor conditions.
- Eftekhari et al (2003) implemented and tested fuzzy controllers for naturally ventilated buildings. The controllers have different numbers of membership functions and rules and were tested simultaneously. The inputs to the controller were indoor and outdoor temperatures, wind velocity and rain direction. The validation of the controllers is performed in the test room by measuring the temperature distribution inside the room with no control action. The data are then compared to the open-loop test results of the controller. The indoor temperatures were in a range between 20 and 27°C and the overall performance is improved.
- Da Graca et al (2007) developed a multi-criteria methodology for evaluation of the environmental comfort parameters of school buildings during preliminary design. Fuzzy grading of thermal comfort, acoustic comfort, natural lighting and functionality was developed. This study presents a good set of comfort variables for future

designs, based on simple relations between form, orientation and location of openings.

- Lah et al (2005) designed a fuzzy control system for the regulation of a movable shade roller blind, targeting the harmonization of thermal and optical flows. They aimed to combine minimization of energy use with comfortable living and working conditions. The control algorithm consists of thermal and lighting parts, each one containing conventional and fuzzy controllers. The impact on thermal light behaviour was analysed with adaptable window geometry. The controller is designed and adjusted so as to adjust the inside daylight illumination level with moderate continuous movement of the roller blind. The largest deviations are caused by very changeable solar radiation in short time periods and by changes in the set point illumination profile.

- Ardehali et al (2004) simulated and analysed the performance of fuzzy logic-based tuning of fuzzy PID and fuzzy PSD controllers for a heating and cooling energy system while taking into account the thermal zone air, heat exchanger and thermal zone wall. Cost functions that incorporate energy and comfort costs are developed for comparison purposes. It is concluded that the utilization of fuzzy logic-based rules resulted in lower energy and comfort costs by 24 per cent and 17 per cent as compared to the PID, and by 13 per cent and 8 per cent as compared to the PSD, respectively. It is also shown that the fuzzy PSD controller performs better, as the associated energy and comfort costs for the fuzzy PID are 10 per cent and 2 per cent higher.

- Hagras et al (2003) presented a hierarchical fuzzy system for occupants' comfort. The system is adjusted to the occupants' needs by GAs. The overall algorithm is based on multi-agent technology. The proposed 'agent' is composed by three behaviours (the safety, emergency and economy behaviours) and an adaptable rule set of comfort behaviours that are adapted according to the occupants' actual behaviour. An 'experience bank' is introduced that applies rule bases according to occupant's behaviour. If the rule base extracted from the experience bank is not suitable for the user, the GA starts its search for new consequences for the poorly performing rules. The system interactively learns the optimized rule base for the comfort behaviour in a small set of interactions and produces similar rules to a rule base learned by off-line supervised techniques. The system has comparable results to off-line approaches (for example, Mendel-Wang approach, offline GA and ANFIS).

- Guillemin and Morel (2001) developed a self-adaptive integrated system for building energy and comfort management. The fuzzy expert system consists of a shading device controller, an artificial lighting controller and a heating controller. When the user is present, priority is given to visual comfort, and when absent, priority is given to thermal aspects (heating/cooling energy saving). The models used in the controller are adapted regularly in order to meet the requirements of the building and of the environment. A process of adaptation is performed each night using GAs in order to identify the most appropriate parameters of the controller. The operation of the controller is compared to a conventional controller (no automatic blind control and artificial lighting control, and a proportional controller for heating). The energy consumption of the fuzzy controller is 20–25 per cent less than a conventional controller.

- Additionally, Guillemin and Morel (2002) presented a controller with the users' set points and the weather data of the room as inputs. ANNs are included for the prediction of the room temperature. The thermal comfort level is kept high and the visual comfort is improved by the fuzzy control system. Energy consumption is reduced.

- Alcala et al (2001; 2003) developed fuzzy logic controllers that control HVAC systems concerning energy performance and indoor comfort requirements that are tuned and optimized by GAs. The fitness function characterizes the performance of each tested controller for thermal comfort, indoor air quality, energy consumption and system stability criteria. With the tuning process of GAs, energy consumption is gradually decreased so that an improvement of almost 16 per cent is achieved. If energy consumption continues to decrease, this happens at the expense of stability. The indoor comfort goals are met.

- An integrated control system that incorporates thermal comfort, visual comfort and indoor air quality was studied by Pargfrieder and Jorgl (2002). The aim is to optimize energy consumption and users' comfort simultaneously. A fuzzy controller with adaptive profile, a GA optimized controller and a predictive controller were analysed using simulation analysis.

- Dounis et al (1994) investigated the impact of natural ventilation on the thermal comfort index, assuming the implementation of fuzzy reasoning for visual comfort control. The analysis is based on mathematical models.

- A hierarchical fuzzy logic controller composed of supervisor, coordinator and local levels was applied by Shoureshi et al (1996) to the problem of building temperature and lighting control. The intelligent control structure is designed to take advantage of variable occupancy and utility cost, and building dynamics are used as inputs to an adaptive fuzzy relation that generates local controller set points based on accumulated knowledge of past building usage and behaviour. The supervisor is autonomously optimized with respect to a performance index penalizing energy consumption, occupant discomfort and fuzzy model error. Implementation of the controller in an office building is discussed.

- A fuzzy PID, a fuzzy PD (proportional derivative) and an adaptive fuzzy PD controller were applied to indoor thermal comfort, indoor air quality and visual comfort by Kolokotsa et al (2001). The adaptive fuzzy PD controller ensures lower energy consumption, even under extreme users' preferences. The reduction of the energy consumption, compared to that of the conventional on-off control, is achieved by optimizing the response of the PMV index. Elimination of overshootings and oscillations that contribute to significant increase of energy waste are noticed.

- The controllers' performance was evaluated in Kolokotsa (2003). The fuzzy proportional controller has the lowest energy consumption when is compared to a fuzzy PID, a fuzzy PD and an adaptive fuzzy PD controller. Additionally, the overshootings and steady-state errors are eliminated when the fuzzy PD and adaptive PD controllers are simulated. The analysis results for a one-day simulation and for the same climatic conditions are in Table 2.4.

● A fuzzy controller for the control of the indoor comfort parameters at the building zone level was proposed by Kolokotsa et al (2002). The occupants' preferences are inserted into the fuzzy controller and GA optimization is applied to properly shift the membership functions of the fuzzy controller in order to satisfy the occupants' preferences while minimizing energy consumption. The energy consumption before and after GA optimization is analysed and the steady-state error is reduced after GA.

● Kolokotsa et al (2005) tested a fuzzy controller in two real installations in Athens and Crete, Greece. The reduction of energy consumption is up to 20 per cent for heating and cooling, and 50–70 per cent for lighting (Table 2.5).

TABLE 2.4 Performance of the zone-level fuzzy controllers for one-day simulation

		FUZZY P	FUZZY PID	FUZZY PI	FUZZY PD	ADAPTIVE PD
Energy consumption (kWh/m²)	Heating	2.89	3.18	3.18	3.18	3
	Electric lighting	0.56	0.63	0.63	0.63	0.78
Response performance	Overshooting (%) PMV	0	1	1	1	1
	CO_2	0	20	20	0	0
	Illuminance	0	0	0	0	0
	Steady-state error PMV	0.1	0	0	0	0
	CO_2 (ppm)	1–2	0	0	0	0
	Illuminance (lux)	50	0	0	0	0

Source: Kolokotsa (2003)

CONCLUSIONS

In the previous sections the application of fuzzy logic in buildings energy and indoor environment research was analysed. In this final section the major observations and outcomes of the analysis are summarized.

Early research focused on the improvement of thermal comfort either directly (by the regulation of the PMV index, the operative temperature, humidity and so on) or indirectly by the improvement of the HVAC systems operation. Thus, in 90 per cent of the research work analysed, thermal comfort and HVAC operation for the improvement of thermal comfort prevail. Indoor air quality and visual comfort are mainly investigated in combination with their impact on thermal comfort and vice versa. Fuzzy logic is also

TABLE 2.5 Evaluation of the energy conservation for the buildings in Crete and Athens, Greece

SUMMER PERIOD – COOLING			
CRETE BUILDING		**ATHENS BUILDING**	
Fuzzy Controller	On-Off Controller	Fuzzy Controller	On-Off Controller
4580kWh	5324kWh	75kWh	123kWh
31.8kWh/m²	37kWh/m²	1.3kWh/m²	2.2kWh/m²
Estimated energy saving 14%		Estimated energy saving 38%	
WINTER PERIOD – HEATING			
CRETE BUILDING		**ATHENS BUILDING**	
Fuzzy Controller	On-Off Controller	Fuzzy Controller	On-Off Controller
3786kWh	5151kWh	1866kWh	2351kWh
26.3kWh/m²	35.8kWh/m²	33.2kWh/m²	41.8kWh/m²
Estimated energy saving 26.5%		Estimated energy saving 20.6%	
ANNUAL PERIOD – COOLING AND HEATING			
CRETE BUILDING		**ATHENS BUILDING**	
Fuzzy Controller	On-Off Controller	Fuzzy Controller	On-Off Controller
8367kWh	10475kWh	1941kWh	2474kWh
58.1kWh/m²	72.7kWh/m²	34.5kWh/m²	44.0kWh/m²
Estimated energy saving 20.1 %		Estimated energy saving 21.5%	
ANNUAL PERIOD – LIGHTING			
CRETE BUILDING		**ATHENS BUILDING**	
Fuzzy Controller	No control	Fuzzy Controller	No control
1045.6kWh	4419.4kWh	3175kWh	5639kWh
7.3kWh/m²	30.7kWh/m²	56.4kWh/m²	100.2kWh/m²
Estimated energy saving 76.3 %		Estimated energy saving 43.7%	
ANNUAL PERIOD – TOTAL ENERGY CONSUMPTION			
CRETE BUILDING		**ATHENS BUILDING**	
Fuzzy Controller	On-Off Controller	Fuzzy Controller	On-Off Controller
9412.6kWh	14894.4kWh	5116kWh	8113kWh
65.4kWh/m²	103.4kWh/m²	91kWh/m²	144.2kWh/m²
Estimated energy saving 36.8 %		Estimated energy saving 36.9%	

Source: Kolokotsa et al (2005)

Table 2.6 gives on overview of the above mentioned research.

TABLE 2.6 Summary of fuzzy logic in hybrid applications

APPLICATION	AI TECHNIQUES	PERFORMANCE AND REMARKS	REFERENCES
Thermal comfort and indoor air quality	● Optimization using fuzzy and GA ● Fuzzy supervisory techniques	10–15% reduction of the energy consumption. Better stability and reduction of the steady-state errors. Indoor thermal comfort within acceptable limits	Eftekhari and Marjanovic, 2003; Eftekhari et al, 2003; Shepherd and Batty, 2003; Alcala et al, 2005
Thermal comfort and visual comfort	● Fuzzy controller ● Self-adapting fuzzy ● Weather prediction in combination fuzzy logic	Up to 25% energy conservation compared to conventional control	Shoureshi et al, 1996; Guillemin and Morel, 2001; 2002; Lah et al, 2005
HVAC and indoor comfort	● Fuzzy PID and PSD ● Fuzzy optimized by GA	Lower energy and comfort costs by up to 24% and 17% compared to conventional control	Alcala et al, 2001; 2003; Ardehali et al, 2004
Thermal, visual comfort and indoor air quality (IAQ)	● Adaptive fuzzy control ● Hierarchical fuzzy system ● GA optimized fuzzy control ● Fuzzy PID, PD and adaptive PD	Eliminations of overshootings and oscillations. Reduction of the steady-state error. Up to 20% reduction in energy consumption compared to conventional control. The reduction of electric lighting energy consumption is higher	Dounis et al, 1994; Kolokotsa et al, 2001; 2002; 2005; Pargfreider and Jorgl, 2002; Hagras et al, 2003; Kolokotsa, 2003; Da Graca et al, 2007

extensively applied in combination with neural networks and GAs for the prediction of the behaviour of HVAC systems in order to improve the performance of either conventional or advanced control systems.

It is unquestionable that fuzzy logic either in stand-alone applications or in combination with other AI technologies significantly improves the indoor comfort and contributes in the minimization of energy consumption. However, conventional methods – for example, set-back controls and economizers (Mathews et al, 2001) – still dominate the researchers' interests, as well as the building automation market and are, to a lesser or greater extent, effective. The main disadvantage of the application of fuzzy logic is the need for expert knowledge for the controller's development. This makes the application of a fuzzy controller dependent on expert experience for development and tuning. Therefore implementation of fuzzy logic may not be as straightforward as binary logic is.

More effort and investigation should be focused at both research and market levels. In almost all the research, the controllers developed are either tested by simulation analysis or by limited experimental analysis and under certain conditions. Extensive testing is needed to assist key decision-makers and end-users to realize the benefits of adaptive and fuzzy logic controls. Fuzzy controllers' stability in buildings' energy and indoor management should be analysed further and possibly various stability criteria could be compared (Kandel et al, 1999).

Lighting control needs further investigation especially when the controlled variable is indoor illuminance. Significant oscillations can appear due to the high disturbances caused by variations in outdoor illuminance. This is a significant point as control of indoor illuminance can significantly reduce energy consumption for electric lighting, as mentioned earlier. The need of extra sensors in order to evaluate glare and other visual comfort parameters is a significant drawback for such applications. The work performed so far is quite significant; however, there are more problems of visual comfort to be solved.

Demand control of indoor air quality cannot be satisfactory and the system lacks reliability. The necessity of sensors limits the possibilities of expansion at the present time. Predictive and adaptive control needs more effort for results accuracy under various conditions.

At this point in time, the information that the market has about fuzzy applications in building automation is 'fuzzy'. From the point of view of installers, contractors, energy managers and so on, the complicated nature of these controls makes installation, troubleshooting and servicing more difficult relative to conventional systems (Westphalen et al, 2003). Simplification, reduction of rules, parameters and extensive testing are needed in order to make fuzzy systems attractive to the building automation market. Adequate and robust documentation of how AI in buildings can contribute to comfort and energy conservation is needed. Intellectual property rights need to be investigated and consolidated. Certification procedures should be followed for mature products. Last but not least, sufficient training of all involved actors should be provided.

AUTHOR CONTACT DETAILS

D. Kolokotsa, Technical Educational Institute of Crete, Department of Natural Resources and Environment, Chania, Crete, Greece
Tel: +30 28210 23017; fax: +30 28210 23003; email: kolokotsa@chania.teicrete.gr

REFERENCES

Alcala, R., Casillas, J., Castro, J. L., Gonzalez, A. and Herrera, F. (2001) 'A multicriteria genetic tuning for fuzzy logic controllers', *Mathware and Soft Computing*, vol 8, no 2, pp179–201

Alcala, R., Benitez, J. M., Cassillas, J., Cordon, O. and Perez, R. (2003) 'Fuzzy control of HVAC systems optimized by genetic algorithms', *Applied Intelligence*, vol 18, pp155–177

Alcala, R., Casillas, J., Cordon, O., Gonzalez, A. and Herrera, F. (2005) 'A genetic rule weighting and selection process for fuzzy control of heating, ventilating and air conditioning systems', *Engineering Applications of Artificial Intelligence*, vol 18, pp279–296

Allard, F., Santamouris, M., Alvarez, S., Daskalaki, E., Guarraccino, G., Maldonado E., Sciuto, S. and Vandaele, L. (1998) *Natural Ventilation in Buildings*. James and James, London

Angelov, P. (1999) 'A fuzzy approach to building thermal systems optimization', *Proceedings of the 8th IFSA World Congress*, Taipei, Taiwan, 17–20 August, vol 2, pp423–426

Angelov, P., Hanby, V., Buswell, R. and Wright, J. (2000a) 'A methodology for modelling HVAC components using evolving fuzzy rules', *IEEE International Conference on Industrial Engineering, Control and Instrumentation*, IECON-2000, 22–28 October 2000, Nagoya, Japan, pp247–252

Angelov, P., Hanby, V. and Wright, J. (2000b) 'HVAC systems simulation: A self-structuring fuzzy rule-based approach', *International Journal of Architectural Sciences*, vol 1, no 1, pp49–58

Angelov, P., Zhang, Y., Wright, J., Buswell, R. and Hanby, V. (2003) 'Automatic design synthesis and optimization of component-based systems by evolutionary algorithms', *Lecture Notes in Computer Science 2724 Genetic and Evolutionary Computation – GECCO 2003*, Springer-Verlag, Berlin, Germany, pp1938–1950

Angelov, P., Zhang, Y. and Wright, J. (2005) 'Automatic design generation of component-based systems using GA and fuzzy optimisation', *Proceedings of the 1st International Workshop on Genetic Fuzzy Systems*, Granada, Spain, pp95–100

Ardehali, M. M., Saboori, M. and Teshnelab, M. (2004) 'Numerical simulation and analysis of fuzzy PID and PSD control methodologies as dynamic energy efficiency measures', *Energy Conversion and Management*, vol 45, pp1981–1992

Ari, S., Cosden, I. A., Khalifa, H. E., Dannenhoffer, J. F., Wilcoxen, P. and Isik, C. (2005) 'Constrained fuzzy logic approximation for indoor comfort and energy optimization', *Annual Meeting of the North American Fuzzy Information Processing Society NAFIPS*, Proceedings, IEEE, Ann Arbor, Michigan

ASHRAE (Standard 62-1999) (1999) *Ventilation for Acceptable Indoor Air Quality*, American Society of Heating, Refrigeration and Air-conditioning Engineers, Atlanta

Bailey, M. B. and Curtiss, P. S. (2001) 'Neural network modelling and control applications in building mechanical systems', *Proceedings from Chartered Institution of Building Services Engineers National Conference*, Regent College, London

Bellas-Vellidis, I., Argiriou, A., Balaras, C. A. and Kontoyannidis, S. (1998) 'Predicting energy demand of single solar houses using artificial neural networks', *Proceedings of the 6th European Congress on Intelligent Techniques and Soft Computing*, Aachen, Germany, pp873–877

Calvino, F., Gennusa, M. L., Rizzo, G. and Scaccianoce, G. (2004) 'The control of indoor thermal comfort conditions: Introducing a fuzzy adaptive controller', *Energy and Buildings*, vol 36, pp97–102

Chao, K., Gates, R. S. and Sigrimis, N. (2000) 'Fuzzy logic controller design for staged heating and ventilating systems', *Transactions of the American Society of Agricultural Engineers*, vol 43, no 6, pp1885–1894

Chen, K. T., Rys, M. J. and Lee, E. S. (2006a) 'Modeling of thermal comfort in air conditioned rooms by fuzzy regression analysis', *Mathematical and Computer Modeling*, vol 43, no 7–8, pp809–819

Chen, K. T., Jiao, Y. and Lee, E. S. (2006b) 'Fuzzy adaptive networks in thermal comfort', *Applied Mathematics Letters*, vol 19, pp420–426

Chu, C. M., Jong, T. L. and Huang, Y. W. (2005) 'Thermal comfort control on multi-room fan coil unit system using LEE-based fuzzy logic', *Energy Conversion and Management*, vol 46, pp1579–1593

CIBSE (Chartered Institution of Building Services Engineers) (2004) *Code for Interior Lighting*, CIBSE, UK

Curtiss, P. S., Kreider, J. F. and Brandelmuehl, M. J. (1993) 'Adaptive control of HVAC processes using predictive neural networks', *ASHRAE Transactions*, vol 99, pp496–504

Da Graca, V. A. C., Kowaltowski, D. C. C. K. and Petreche, J. R. D. (2007) 'An evaluation method for school building design at the preliminary phase with optimisation of aspects of environmental comfort for the school system of the State Sao Paulo in Brazil', *Building and Environment*, vol 42, no 2, pp984–999

Dexter, A. L. and Benouarets, M. (1997) 'Model-based fault diagnosis using fuzzy matching', *IEEE Transactions on Systems, Man, and Cybernetics*, part A, vol 27, no 5, pp673–682

Dexter, A. L. and Trewhella, D. W. (1990) 'Building control systems: Fuzzy rule based approach to performance assessment', *Building Services Research and Technology*, vol 11, no 4, pp115–124

Dounis, A. I., Santamouris, M. and Lefas, C. C. (1992) 'Implementation of artificial intelligence techniques in thermal comfort control for passive solar buildings', *Energy Conversion and Management*, vol 33, no 3, pp175–182

Dounis, A. I., Santamouris, M. and Lefas, C.C. (1993) 'Building visual comfort control with fuzzy reasoning', *Energy Conversion and Management*, vol 34, no 1, pp17–28

Dounis, A. I., Santamouris, M., Lefas, C. C. and Manolakis, D. E. (1994) 'Thermal-comfort degradation by a visual comfort fuzzy-reasoning machine under natural ventilation', *Applied Energy*, vol 48, no 2, pp115–130

Dounis, A. I., Lefas, C. C. and Argiriou, A. (1995a) 'Knowledge base versus classic control for solar building design', *Applied Energy*, vol 50, no 4, pp281–292

Dounis, A. I., Manolakis, D. E. and Argiriou, A. (1995b) 'Fuzzy rule-based approach to achieve visual comfort conditions', *International Journal of Systems Science*, vol 26, no 7, pp1349–1361

Dounis, A. I., Santamouris, M., Lefas, C. C. and Argiriou, A. (1995c) 'Design of a fuzzy set environment comfort system', *Energy and Buildings*, vol 22, no 1, pp81–87

Dounis, A. I., Bruant, M., Guaraccino, G., Michel, P. and Santamouris, M. (1996a) 'Indoor air-quality control by a fuzzy-reasoning machine in naturally ventilated buildings', *Applied Energy*, vol 54, no 1, pp11–28

Dounis, A. I., Bruant, M., Santamouris, M., Guaraccino, G. and Michel, P. (1996b) 'Comparison of conventional and fuzzy control of indoor air quality in buildings', *Journal of Intelligent and Fuzzy Systems*, vol 4, pp131–140

Dounis, A. I. and Manolakis, D. E. (2001) 'Design of a fuzzy system for living space thermal-comfort regulation', *Applied Energy*, vol 69, pp119–144

Dubois, D. and Prade, H. (1980) *Fuzzy Sets and Systems: Theory and Application*, Academic Press, New York

Eftekhari, M. M. and Marjanovic, L. D. (2003) 'Application of fuzzy control in naturally ventilated buildings for summer conditions', *Energy and Buildings*, vol 35, pp645–655

Eftekhari, M. M., Marjanovic, L. D. and Angelov, P. (2003) 'Design and performance of a rule-based controller in a naturally ventilated room', *Computers in Industry*, vol 51, pp299–326

Egilegor, B., Uribe, J. P., Arregi, G., Pradilla, E. and Susperregi, L. (1997) 'A fuzzy control adapted by a neural network to maintain a dwelling within thermal comfort', *5th International IBPSA Conference, Building Simulation*, Prague, vol II. pp87–94

Fanger, P. O. (1970) *Thermal Analysis and Applications in Environmental Engineering*, McGraw Hill, New York

Feriadi, H. and Hien, W. N. (2003) 'Modelling thermal comfort for tropics using fuzzy logic, *8th International IBPSA Conference*, Eindhoven, August 11–14

Fraisse, G., Virgone, J. and Brau, J. (1997a) 'An analysis of the performance of different intermittent heating controllers and an evaluation of comfort and energy consumption', *Heating Ventilation Air Conditioning and Research*, vol 3, no 4, pp369–386

Fraisse, G., Virgone, J. and Roux, J. J. (1997b) 'Thermal comfort of discontinuously occupied building using a classical and a fuzzy logic approach', *Energy and Buildings*, vol 26, pp303–316

Fraisse, G., Virgone, J. and Yezou, R. (1999) 'A numerical comparison of different methods for optimizing heating-restart time in intermittently occupied buildings', *Applied Energy*, vol 62, pp125–140

Ghiaus, C. (2001) 'Fuzzy model and control of a fan-coil', *Energy and Buildings*, vol 33, pp545–551

Gouda, M. M., Danaher, S. and Underwood, C. P. (2000) 'Fuzzy logic controller versus conventional PID controller for controlling indoor temperature of a building space', *Computer Aided Control System Design Conference*, Salford University, Salford

Gouda, M. M, Danaher, S. and Underwood, C. P. (2001) 'Thermal comfort based fuzzy logic controller', *Building Services Engineering Research and Technology*, vol 22, no 4, pp237–253

Gouda, M. M., Danaher, S. and Underwood, C. P. (2006) 'Quasi-adaptive fuzzy heating control of solar buildings', *Building and Environment*, vol 41, no 12, pp1881–1891

Guillemin, A. and Molteni, S. (2002) 'An energy-efficient controller for shading devices self-adapting to the user wishes', *Building and Environment* , vol 37, pp1091–1097

Guillemin, A. and Morel, N. (2001) 'An innovative lighting controller integrated in a self-adaptive building control system', *Energy and Buildings*, vol 33, no 5, pp477–487

Guillemin, A. and Morel, N. (2002) 'Experimental results of a self-adaptive integrated control system in buildings: A pilot study', *Solar Energy*, vol 72, no 5, pp397–403

Hagras, H., Callaghan, V., Colley, M. and Clarke, G. (2003) 'A hierarchical fuzzy–genetic multi-agent architecture for intelligent buildings online learning, adaptation and control', *Information Sciences*, vol 150, pp33–57

Hamdi, M., Lachiver, G. and Michaud, F. (1999) 'A new predictive thermal sensation index of human response', *Energy and Buildings*, vol 29, pp167–178

Han, Y., Xiu, L., Wang, Z., Chen, Q. and Tang, S. (1997) 'Artificial neural networks controlled fast valving in power generation plant', *IEEE Transactions on Neural Networks*, vol 8, pp373–389

He, M., Cai, W. J. and Li, S. Y. (2005) 'Multiple fuzzy model-based temperature predictive control for HVAC systems', *Information Sciences*, vol 169, pp155–174

Huang, W. and Lam, H. N. (1997) 'Using genetic algorithms to optimize controller parameters for HVAC systems', *Energy and Buildings*, vol 26, no 3, pp277–282

Huang, S. H. and Nelson, R. M. (1994a) 'Rule development and adjustment strategies of a fuzzy logic controller for HVAC system: Part one, analysis', *ASHRAE Transactions*, vol 100, no 1, pp841–850

Huang, S. H. and Nelson, R. M. (1994b) 'Rule development and adjustment strategies of a fuzzy logic controller for HVAC system: Part two, experiment', *ASHRAE Transactions*, vol 100, no 1, pp851–856

Huang, S. H. and Nelson, R. M. (1999) 'Development of a self-tuning fuzzy logic controller', *ASHRAE Transactions*, vol 105

Jana, C. and Chattopadhyay, R. N. (2004) 'Block level energy planning for domestic lighting: A multi-objective fuzzy linear programming approach', *Energy*, vol 29, no 11, pp1819–1829

Kalogirou, S. A. (2000) 'Applications of artificial neural networks for energy systems', *Applied Energy*, vol 67 pp17–35

Kandel, A., Luo, Y. and Zhang, Y. Q. (1999) 'Stability analysis of fuzzy control systems', *Fuzzy Sets and Systems*, vol 105, pp33–48

Karatasou, S., Santamouris, M. and Geros, V. (2006) 'Modelling and predicting building's energy use with artificial neural networks: Methods and results', *Energy and Buildings*, vol 38, no 8, pp949–958

Khotazad, A., Afkhami-Rohani, R., Lu, T., Abaye, A., Davis, M. and Maratukulam, D. J. (1997) 'ANNTSLF-a neural network based electric load forecasting system', *IEEE Transactions on Neural Networks*, vol 8, pp835–846

Kolokotsa, D. (2003) 'Comparison of the performance of fuzzy controllers for the management of the indoor environment', *Building and Environment*, vol 38, no 12, pp1439–1450

Kolokotsa, D., Tsiavos, D., Stavrakakis, G., Kalaitzakis, K. and Antonidakis, E. (2001) 'Advanced fuzzy logic controllers design and evaluation for buildings' occupants thermal – visual comfort and indoor air quality satisfaction', *Energy and Buildings*, vol 33, no 6, pp531–543

Kolokotsa, D., Stavrakakis, G., Kalaitzakis, K. and Agoris, D. (2002) 'Genetic algorithms optimised fuzzy controller for the indoor environmental management in buildings implemented using PLC and local operating networks', *Engineering Applications of Artificial Intelligence*, vol 15, no 5, pp417–428

Kolokotsa, D., Stavrakakis, G. and Agoris, D. (2003) 'Optimized Fuzzy Controller for Indoor Comfort Control and Energy Management in Buildings Implemented with PLC and Local Operating Networks (LON) Technology', *Computer Science and Technology Conference*, Cancun, Mexico

Kolokotsa, D., Niachou, K., Geros, V., Kalaitzakis, K., Stavrakakis, G. and Santamouris, M. (2005) 'Implementation of an integrated indoor environment and energy management system', *Energy and Buildings*, vol 37, pp93–99

Kolokotsa, D., Pouliezos, A. and Stavrakakis, G. (2006) 'Sensor fault detection in building energy management systems', *International Conference on Intelligent Systems and Knowledge Engineering (ISKE2006)*, Shanghai, China

Krarti, M. (2003) 'An overview of artificial intelligence based methods for building energy systems', *Journal of Solar Engineering*, vol 125, pp331–340

Kreider, J. F. (1995) 'Neural networks applied to building energy studies', in *System Identification Applied to Building Performance Data*, Joint Research Center, Ispra, pp233–251

Kreider, J. F., Curtiss, P. and Rabl, A. (2002) *Heating and Cooling for buildings, Design for efficiency*, McGraw-Hill, New York

Kubota, N., Hashimoto, S., Kojima, F. and Taniguch, M. K. (2000) 'GP-preprocessed fuzzy inference for the energy load prediction', *IEEE Proceedings of the 2000 Congress on Evolutional Computation*, pp1–6

Kurian, C. P., Kuriachan, S., Bhat, J. and Aithal, R. S. (2005) 'An adaptive neuro-fuzzy model for the prediction and control of light in integrated lighting schemes', *Lighting Research and Technology*, vol 37, no 4, pp343–352

Lah, M. T., Zupancic, B. and Krainer, A. (2005) 'Fuzzy control for the illumination and temperature comfort in a test chamber', *Building and Environment*, vol 40, pp1626–1637

Lah, M. T., Zupancic, B., Peternelj, J. and Krainer, A. (2006) 'Daylight illuminance control with fuzzy logic', *Solar Energy*, vol 80, pp307–321

Lea, R. N., Dohmann, E., Prebilsky, W. and Jani, Y. (1996) 'HVAC fuzzy logic zone control system and performance results', *IEEE International Conference on Fuzzy Systems*, vol 3, pp2175–2180

Levermore, G. J. (2000) *Building Energy Management Systems: An Application to Heating, Natural Ventilation, Lighting and Occupant Satisfaction*, Taylor and Francis, London

Liao, Z. (2005) 'A neural-fuzzy based inferential sensor for improving the control of boilers in space heating systems', *Lecture Notes in Computer Science*, vol 3612, no 3, pp1205–1215

Liu, X. F. and Dexter, A. (2001) 'Fault-tolerant supervisory control of VAV air-conditioning systems', *Energy and Buildings*, vol 33, pp379–389

Loveday, D. L. and Virk, G. S. (1992) 'Artificial intelligence for buildings', *Applied Energy*, vol 41, no 3, pp201–221

Lu, L., Cai, W., Xie, L., Li, S. and Soh, Y. C. (2005) 'HVAC system optimization: In-building section', *Energy and Buildings*, vol 37, pp11–22

Maor, I. and Reddy, T. A. (2003) 'Literature review of artificial intelligence and knowledge-based expert systems in buildings and HVAC&R system design', *ASHRAE Winter Meetings CD, Technical and Symposium Papers*, vol 2003, pp11–24

Mathews, E. H., Botha, C. P., Arndt, D. C. and Malan, A. (2001) 'HVAC control strategies to enhance comfort and minimise energy usage', *Energy and Buildings*, vol 33, no 8, pp853–863

Michalik, G., Khan, M. E., Bonwick, W. J. and Mielczarski, W. (1997) 'Structural modelling of energy demand in the residential sector: 2. The use of linguistic variables to include uncertainty of customers' behaviour', *Energy*, vol 22, pp949–958

Ngo, D. and Dexter, A. L. (1998) 'Automatic commissioning of air-conditioning plant', *IEE Conference Publication*, no 455, pp1694–1699

Ngo, D. and Dexter, A. L. (1999) 'Robust model-based approach to diagnosing faults in air-handling units', *ASHRAE Transactions*, vol 105, pp1078–1086

Nilsson, N. (1998) 'Artificial intelligence, a new synthesis', *The Morgan and Kaufman Series in Artificial Intelligence*, Morgan and Kaufmann, San Francisco, CA

Olesen, B. W. and Parsons, K. C. (2002) 'Introduction to thermal comfort standards and to new version of EN ISO 7730', *Energy and Buildings*, vol 34, pp537–548

Pargfrieder, J. and Jorgl, H. P. (2002) 'An integrated control system for optimizing the energy consumption and user comfort in buildings', *IEEE International Symposium on Computer Aided Control System Design Proceedings*, pp127–132

Piao, Y. G., Zhang, H. G. and Zeungnam, B. (1998) 'A simple fuzzy adaptive control method and application in HVAC', *IEEE International Conference on Fuzzy Systems*, vol 1, pp528–532

Rahmati, A., Rashidi, F. and Rashidi, M. (2003) 'A hybrid fuzzy logic and PID controller for control of nonlinear HVAC systems', *IEEE International Conference on Systems, Man and Cybernetics*, vol 3, pp2249–2254

Reyes-Garcia, C. A. and Corona, E. (2003) 'Implementing fuzzy expert system for intelligent buildings', *Proceedings of the ACM Symposium on Applied Computing*, pp9–13

Shepherd, A. B. and Batty, W. J. (2003) 'Fuzzy control strategies to provide cost and energy efficient high quality indoor environments in buildings with high occupant densities', *Building Services Engineering Research Technology*, vol 24, no 1, pp35–45

Shimizu, Y. and Jindo, T. (1995) 'Fuzzy logic analysis method for evaluating human sensitivities', *International Journal of Industrial Ergonomics*, vol 15, pp39–47

Shoureshi, R., Brackney, L. and DeRoo, B. (1996) 'Fuzzy-based energy management system for large buildings', *American Society of Mechanical Engineers, Dynamic Systems and Control Division*, vol 58, pp797–804

Singh, J., Singh, N. and Sharma, J. K. (2005) 'Complexity reduction in fuzzy modelling of fan-coil unit of HVAC system', *Journal of Scientific and Industrial Research*, vol 64, no 6, pp420–425

Skrjanc, I., Zupancic, B., Furlanb, B. and Krainerb, A. (2001) 'Theoretical and experimental fuzzy modelling of building thermal dynamic response', *Building and Environment*, vol 36, pp1023–1038

So, A. T. P., Chan, W. L. and Tse, W. L. (1997) 'Self learning fuzzy air handling system controller', *Building Services Engineering Research Technology*, vol 18, no 2, pp99–108

Sousa, J. M., Babuska, R. and Verbruggen, H. B.(1997) 'Fuzzy predictive control applied to an air conditioning system', *Control Engineering Practice*, vol 10, no 5, pp1395–1406

Sulaiman, F., Ahmad, A. and Ahmed, A. Z. (2005) 'Fuzzy logic algorithm for automated dimming control used in passive optical fiber daylighting system for energy savings', *WSEAS Transactions on Electronics*, vol 2, no 4, pp189–196

Wen, X. J. and Zhao, R. Y. (1998) 'Fuzzy comprehensive evaluation of thermal sensation in dynamic thermal environment', *Qinghua Daxue Xuebao (Journal of Tinghua University)*, vol 38, pp23–27

Westphalen, D., Roth, K. W. and Brodrick, J. (2003) 'Fuzzy logic for controls', *ASHRAE Journal*, vol 45, pp31–32

Wright, J. A., Loosemore H. A. and Farmani, R. (2002) 'Optimization of building thermal design and control by multi-criterion genetic algorithm', *Energy and Buildings*, vol 34, no 9, pp959–972

Yonezawa, K., Yamada, F. and Wada, Y. (2000) 'Comfort air-conditioning control for building energy saving', *IECON, 26th Annual Conference of the IEEE*, vol 3, pp1737–1742

Zadeh, L. (1973) 'Outline of a new approach to the analysis of complex systems and decision processes', *IEEE Transactions on Systems, Man and Cybernetics*, pp28–44

Zampolli, S., Elmi, I., Ahmed, F., Passini, M., Cardinali, G. C., Nicoletti, S. and Dori, L. (2004) 'An electronic nose based on solid state sensor arrays for low-cost indoor air quality monitoring applications', *Sensors and Actuators*, vol B101, pp39–46

Zhang, J., Ou, J. and Sun, D. (2003a) 'Study on fuzzy control for HVAC systems, Part One: FFSI and development of single-chip fuzzy controller', *ASHRAE Transactions: Research*, vol 109, pp27–35

Zhang, J., Ou, J. and Sun, D. (2003b) 'Study on fuzzy control for HVAC Systems, Part Two: Experiments of FFSI in testing, room dynamic thermal system', *ASHRAE Transactions: Research*, vol 109, pp36–43

Zheng, W. and Xu, H. (2004) 'Design and application of self-regulating fuzzy controller based on qualitative and quantitative variables', *Proceedings of the World Congress on Intelligent Control and Automation (WCICA)*, vol 3, pp2472–2475

3

Field Studies of Indoor Thermal Comfort and the Progress of the Adaptive Approach

Michael A. Humphreys, J. Fergus Nicol and Iftikhar A. Raja

Abstract

The experimental basis of the adaptive approach to thermal comfort is the field study. Since the publication of Bedford's study in 1936, many researchers have collected data on people's thermal comfort in everyday conditions. The principal assemblages of data are those of Humphreys (1975; 1978; 1981), de Dear et al (1997), de Dear (1998), de Dear and Brager (1998) and McCartney and Nicol (2001). In addition there exist numerous studies not included in any assemblage. The data come from a variety of climates and countries, from people in buildings that are heated or cooled mechanically, and from buildings operating without either heating or cooling. This paper compares the more recent data with the early patterns revealed by Humphreys' 1975–1981 analyses of the worldwide data. The comparison is made for indices of thermal comfort, mean warmth response, temperatures found to be comfortable, sensitivity to temperature change, interpersonal variation, distribution of comfort with temperature, and the influence of outdoor temperature in the presence or absence of heating and cooling. The empirical findings are compared with predictions from the predicted mean vote (PMV) model of comfort. The role of the adaptive approach in the formation of standards and guidelines is discussed.

■ *Keywords* – thermal comfort; field studies; climate; meta-analysis; adaptive approach; standards

INTRODUCTION
RESEARCH METHODS

Thermal comfort may be approached from the standpoint of thermal physiology. This approach asks what body-states people find comfortable at various levels of activity, establishes the heat and moisture transfer properties of clothing, and evaluates the effects of the thermal variables (air temperature, radiation exchange, air movement and humidity). The research is commonly conducted in climate-controlled rooms, with subjects in standard clothing and performing standard tasks. The strength of the approach is the possibility of comprehensive measurement in controlled conditions, and the use of sound experimental design. The resulting models of human thermal response are used to assess the effect of any proposed environment and clothing ensemble. The best known model is Fanger's PMV (Fanger, 1970), now in International Standard ISO 7730 (ISO, 2005). It predicts the mean thermal sensation of a group of people on a scale from cold (–3) through neutral (0) to hot (+3), together with the predicted percentage of people dissatisfied (PPD) with the environment. The PMV/PPD model is relatively simple and applies to steady-state conditions. A more complex model capable of predicting responses to changing environments has recently been constructed by Fiala (Fiala, 1998; Fiala et al, 1999). The findings of laboratory-based research and its resulting mathematical models require validation in the conditions of everyday life.

Thermal comfort may also be approached from the standpoint of human adaptation. This adaptive approach investigates the dynamic relation between people and their everyday environments, paying attention to the 'adaptations' people make to their clothing and to their thermal environment to secure comfort. It sees thermal comfort as part of a self-regulating system (Nicol and Humphreys, 1973). Because it concerns the whole range of actions people take to ensure their comfort, the adaptive approach touches on many topics including climatology, the design and construction of buildings, the provision and use of thermal controls, the history and sociology of clothing and the influences of culture, together with human thermal physiology. It therefore encompasses all those aspects of thermal comfort studied in the laboratory, since these researches are themselves a response to the desire for comfort. The principal research method is the field survey. People are asked for their response to their thermal environment, which is measured at the time. Notes of the clothing and of the activity may be kept, from which the thermal insulation of the ensembles and the metabolic rates of the people can be estimated. The opening or closing of windows, the raising or lowering of blinds, and the switching on or off of fans may be noted, together with any other actions that people take to ensure their thermal comfort. Usually no attempt is made by the researcher to control the environment. From such field studies an understanding has developed of how people achieve thermal comfort in daily life, and what environments people typically create or accept in different cultures and climates. It is with the progress of the adaptive approach, and therefore with the accumulated information from thermal comfort field studies, that this review is concerned.

Although the two types of research are complementary, field studies have shown the diversity of the environments that populations find comfortable to be greater than can readily be explained by current physiological models. With the strong likelihood of global warming, and in an era of increasingly expensive fuel, there is powerful incentive to reduce energy use in buildings. If field studies, rather than current physiologically based comfort models, guided the formulation of standards for thermal comfort in buildings, consumption of energy for heating and cooling could be reduced without sacrificing comfort or well-being. Traditional vernacular building styles found comfortable by past generations often cannot meet current physiologically based standards, while they may meet a standard based on the adaptive approach. Well-formulated and appropriately applied adaptive standards could therefore encourage the development of climate-friendly building designs – a new vernacular architecture (see for example, Nicol and Roaf, 2007).

THE EARLY META-ANALYSIS OF THERMAL COMFORT DATA

It is 30 years since the first meta-analysis of thermal comfort field studies worldwide (Humphreys 1975). The analysis brought together data from some 30 field studies of thermal comfort, each broadly following the pattern established by Thomas Bedford (Bedford, 1936). He had measured the thermal environments of some 2500 workers in light industry in the UK winter; the measurements including the temperature of the air, the mean radiant temperature of the surrounding surfaces, the speed of the air movement and its humidity. He asked each person how warm or cool they felt and classified their answers on a seven-category scale of warmth and comfort. The numerous Bedford-pattern field studies that Humphreys collated summarized over 200,000 observations of thermal comfort from Africa, the Americas, Asia, Australasia and Europe. Each of these observations comprised a subjective rating of warmth together with corresponding measurements of the thermal environment. From these data it was possible to compare the effectiveness of various indices of the thermal environment, and the behaviour of various subjective rating scales of warmth and comfort. The temperatures for thermal neutrality or comfort from the various studies were extracted and compared, and related to the temperatures prevailing in the accommodation. The sensitivity of the occupants to changes in room temperature was estimated and distributions of the 'comfort votes' derived, both for adults and for young children. There was a demonstrable disparity between the predictions of the physiological models and the empirical results of the field studies, particularly for hot conditions, and this disparity has been the subject of subsequent research. R. K. Macpherson and A. Auliciems (personal communications, 1975) suggested an extension of the meta-analysis to include the influence of climate, and this further analysis (Humphreys, 1978) evaluated the dependence of comfort conditions on the outdoor temperature. The two parts of the analysis were later drawn together (Humphreys, 1981). These papers have shaped subsequent development of the adaptive approach (see, for example, McIntyre, 1980; Auliciems, 1981; Humphreys, 1995; Humphreys and Nicol, 1998; de Dear and Brager, 1998).

Conclusions arising from the analysis of so large a body of data are not quickly set aside, but over the years sufficient new data accumulate to enable refinements and changes to be made. In this review some of these accumulated data are examined to see how they confirm, modify and develop the conclusions of the meta-analyses of 30 years ago. The adaptive approach has involved many researchers, and their results have been published in the journals of a variety of disciplines, with similar ideas being proposed and developed in somewhat different ways. A truly comprehensive review would be difficult to achieve, as would the accurate attribution of each development. However, we hope we have been able to include the principal developments contributing to the progress of the adaptive model. The review limits itself to thermal comfort indoors. Research into comfort in transitional and outdoor spaces must await a further paper.

FIELD SURVEY DATA SINCE 1975

The collation of data and its analysis are being continually updated, and this review rests on approximately 65,000 observations of thermal comfort and their accompanying measurements of the thermal environment. Many of the more detailed of surveys made

TABLE 3.1 Thermal comfort databases used in this review

SOURCE OF DATA	COUNTRIES (TOWNS)	APPROX. NO. OF OBSERVATIONS	REFERENCES
de Dear	Australia, Canada, Greece, Indonesia, Pakistan, Singapore, Thailand, UK, US	20,000	de Dear et al, 1997; de Dear, 1998
Paktrans	Pakistan (Islamabad, Karachi, Multan, Saidu Sherif, Peshawar)	7000	Nicol et al, 1999
Abdnox	UK (Aberdeen and Oxford)	5000	Raja et al, 1998
SCATs	France, Greece, Portugal, Sweden, UK	4500	McCartney and Nicol, 2001
Oseland	UK (various English towns)	20,000	Oseland, 1998
Kwok	US (Hawaii) (high school students)	3500	Kwok, 1998
Malama	Zambia	4000	Malama et al, 1998
Chan	China (Hong Kong)	2000	Chan et al, 1998
	Total:	66,500	

from 1975 to 1997 were included in the de Dear database, described below. There is also a large study from Pakistan (Nicol et al, 1999) of thermal comfort in five cities in different climate zones, surveyed monthly throughout the year. The SCATs database, described below, is also a year-round study, but of a selection of cities across Europe. Table 3.1 gives brief notes on these and some other important recent field studies. Several other recent surveys currently await inclusion in our analysis.

THE DE DEAR DATABASE

Richard de Dear and those who have worked with him, including Andris Auliciems and Gail Brager, have over many years made important contributions to the development of the adaptive approach to thermal comfort, not least in the assembly and analysis of field data. In the late 1990s de Dear and colleagues collated more than 20,000 sets of data from various researchers to form a database of thermal comfort studies from around the world (Australia, Canada, Greece, Indonesia, Pakistan, Singapore, Thailand, UK and US). We shall call this the de Dear database. To be included in the database the raw data had to be available in electronic form and the records sufficiently complete for values of PMV to be calculated. 'Comfort votes' were on a seven-category rating scale, usually the ASHRAE (American Society of Heating, Refrigeration and Air-conditioning Engineers) scale, but sometimes the Bedford scale, or a semantic differential scale. Some also included responses on a thermal preference scale, usually the McIntyre scale, or Nicol's five-point scale (the scales are shown in Table 3.2). A description of the database can be found in de Dear et al (1997) and de Dear (1998), together with a meta-analysis of the data upon which rests the adaptive extension to the ASHRAE thermal comfort standard (ASHRAE, 2004; de Dear and Brager, 1998).

THE PAKISTAN DATA

The de Dear database included summer and winter data from Pakistan. In five cities six people had provided some 100 responses over a week or so, at home and at work

TABLE 3.2 Scales of warmth and preference

CODE	ASHRAE	BEDFORD	MCINTYRE	NICOL
3	hot	Much too warm		
2	warm	Too warm		Prefer much cooler
1	slightly warm	Comfortably warm	Prefer cooler	Prefer a bit cooler
0	neutral	Comfortable	No change	No change
−1	slightly cool	Comfortably cool	Prefer warmer	Prefer a bit warmer
−2	cool	Too cool		Prefer much warmer
−3	cold	Much too cool		

Semantic differential scale:

Too cold | −3 | −2 | −1 | 0 | 1 | 2 | 3 | Too hot

Comfortable

(a longitudinal experimental design). A large subsequent follow-up study was conducted monthly throughout the year, in the same cities, each of many respondents providing one evaluation in each monthly survey (a transverse experimental design). This is the 'Paktrans' database of Table 3.1.

THE ABERDEEN AND OXFORD DATA

Experimental year-round surveys of design similar to those conducted in Pakistan were undertaken in the UK, (Aberdeen in the north, Oxford in the south) in a selection of air-conditioned and naturally ventilated offices. This is the 'Abdnox' database of Table 3.1.

THE SCATS DATABASE

Research into the European office environment provides a further database. The data come from the SCATs (Smart Controls and Thermal Comfort) project, a Europe-wide survey of the office environment, on which several research teams collaborated. Measurements of the indoor environment were made at selected work areas in 26 office buildings in five European countries (France, Greece, Portugal, Sweden and UK). The participating people were visited monthly, and more than 4500 'desk visits' were made between June 1998 and October 1999. A mobile automated data-acquisition system recorded the thermal environment, the concentration of carbon dioxide in the air, the horizontal illuminance and the sound pressure level at the workstations. Outdoor temperatures were obtained from nearby meteorological stations. While the physical measurements were being recorded, the respondents were interviewed and asked to evaluate the environment at their workstation. The questionnaire included rating scales for their thermal sensation (the ASHRAE scale), their thermal preference (a five-point scale), as well as similar rating and preference scales for air movement, humidity, light, sound and air quality. The scales were in the language of each country, and presented using a flip chart. The year-round duration of the project and the use of identical methods across five countries provided a large database of unique value. A description of the data collection, the instrumentation and some of the findings of the project may be found in Nicol and Sykes (1998), Wilson et al (2000), McCartney and Nicol (2001) and Humphreys et al (2002). In this paper it is only the thermal comfort data that are of interest.

FURTHER DATA

We have also been able to include Oseland's large year-round study of UK office workers (Oseland, 1998), Kwok's study of high school students in Hawaii (Kwok, 1998), Malama's two-season study in dwellings in Zambia (Malamar et al, 1998), and Chan's study of office workers in Hong Kong (Chan et al, 1998). Numerous other studies await inclusion.

COMPARISON WITH THE EARLIER FINDINGS
THERMAL INDICES

A thermal index is a measure that aims to arrange thermal environments according to how warm they feel. The numerous indices then in use* were compared using the Pearson correlation coefficient between the 'comfort votes' of the respondents and the corresponding values of the index. Humphreys concluded that '... simple measurements such as air temperature or globe temperature are often not inferior to the more complex indices in their ability to explain the subjective response of warmth in field studies' (Humphreys, 1975).

Most of those indices have ceased to be used to assess ordinary indoor conditions. Still in use is the simple air temperature (T_a), and at the next level of complexity the operative temperature (T_{op}), a weighted mean of the air temperature and the mean radiant temperature, the weights being in proportion to the convective and the radiant heat-transfer coefficients of the clothed human body. For everyday purposes it is given by the temperature measured by a globe thermometer, but it is not strictly speaking a property of the room environment alone, for the coefficients depend also on the surface properties, activity and orientation of the occupant. Most researchers now use for the globe a table-tennis ball painted grey or matt black, because a sphere of this size has approximately the correct relative response to the air and the radiant temperature at the low air speeds typical of occupied rooms (Humphreys, 1977). Next in complexity is ET*, which replaces Yaglou's original effective temperature (ET). It adds to the operative temperature an allowance for humidity (Gagge et al, 1986). Neither T_{op} nor ET* fully allow for the effect of air movement; they are unaffected by air movement if the air and radiant temperatures are equal. More complex indices include the six main thermal variables: air temperature, thermal radiation, air movement, humidity, clothing insulation and metabolic heat production. Among these are SET (standard effective temperature) (Gagge et al, 1986) and PMV. These indices, derived from the heat exchanges between the human body and the environment, are sometimes called rational indices, to distinguish them from indices derived empirically by, for example, using multivariate statistical analysis to obtain a combination of the thermal variables suitable to predict the

---

* Indices encountered in the 1975 review were: Air Temperature, Equivalent Temperature, Effective Temperature, Corrected Effective Temperature, Globe Temperature, Mean Radiant Temperature, Resultant Temperature, Wet Bulb Temperature, Index Temperature, Dry Kata Cooling Power, Equatorial Comfort Index, Singapore Index, Predicted Four Hour Sweat Rate, Belding-Hatch Heat Stress Index, and the Wet-Bulb-Globe Temperature Index.

TABLE 3.3 Pearson correlation coefficients for the warmth scale and the principal indices

DE DEAR DATABASE					
	TSV	T_A	T_{op}	ET*	SET	PMV
TSV	1	0.514	0.515	0.507	0.430	0.462
T_a	0.514	1	0.996	0.980	0.776	0.834
T_{op}	0.515	0.996	1	0.985	0.776	0.834
ET*	0.507	0.980	0.985	1	0.783	0.827
SET	0.430	0.776	0.776	0.783	1	0.931
PMV	0.462	0.834	0.834	0.827	0.931	1

N=20,468

SCATs DATABASE					
	TSV	T_A	T_{op}	ET*	SET	PMV
TSV	1	0.352	0.333	0.319	0.200	0.249
T_a	0.352	1	0.968	0.912	0.320	0.442
T_{op}	0.333	0.968	1	0.952	0.371	0.488
ET*	0.319	0.912	0.952	1	0.410	0.489
SET	0.200	0.320	0.371	0.410	1	0.848
PMV	0.249	0.442	0.488	0.489	0.848	1

N=4068

'comfort vote'. This latter method of constructing an index goes back to Bedford, but has rarely been used since 1975; the only example we are aware of is Sharma and Ali's carefully constructed Tropical Summer Index (Sharma and Ali, 1986). It seems that the conditions for the valid use of multivariate regression analysis for constructing an index of thermal comfort are, as noted in 1975, often too severe for a reliable index to be derived.

How well do the rational indices correlate with the thermal sensation votes (TSVs) of the respondents? The de Dear database and the SCATs database may both be used for this evaluation, for they include a diversity of indoor environments, thus affording stringent tests of the indices (Table 3.3).

The rational indices do not improve upon the air temperature or the globe temperature. ET* is a little inferior, while SET and PMV are considerably inferior. So the theoretical advantage of including all six thermal variables does not produce a practical benefit; instead there is a deficit. This is attributable jointly to imperfections in their formulaic structure and to error in the measurement of their constituent variables. Figures 3.1 and 3.2 show comparisons of SET and PMV calculated from the values of the six thermal variables. Since the input data are identical for both indices, the scatter arises solely from their different formulations. This comparison reveals considerable imperfection in the formulation of either or both indices. Such effects can cause systematic errors of practical importance in evaluating the environment and these have been quantified for PMV for the de Dear database (Humphreys and Nicol, 2002).

The reason for the relative success of the air temperature is easy to understand. Usually in indoor environments the air temperature and the radiant temperature are not very different. (It seems that air and radiant temperatures are closer in practice than

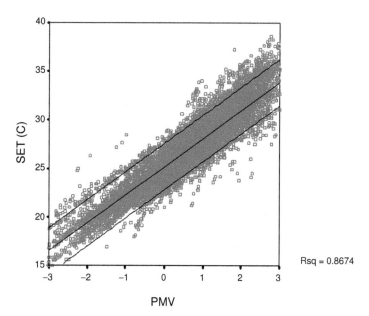

Note: Each point is the calculated value of PMV compared with that of SET for the same thermal environment.
Source: Humphreys and Nicol (2000)

FIGURE 3.1 Scatter plot of PMV and SET for the de Dear data

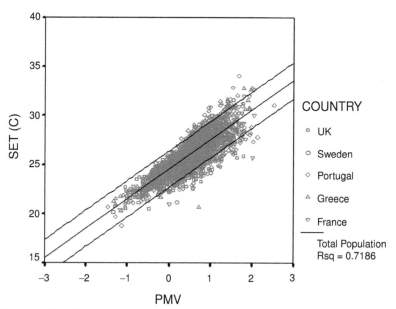

Note: Each point is the calculated value of PMV compared with that of SET for the same thermal environment.
Source: Humphreys and Nicol (2000)

FIGURE 3.2 Scatter plot of PMV and SET for the SCATs data

thermal modelling of an empty room predicts. The presence of furnishings can greatly increase the surface area available for heat exchange with the room air, so reducing the air-radiant temperature difference. The difference also depends on building design, type of heating or cooling, and the outdoor temperature.) It follows that the air temperature and the operative or globe temperature differ little, as is evident from their high correlation (Table 3.3), so either is usually an adequate measure of the temperature-component of the thermal environment. The air movement in a room is often so slight that natural convection prevails at the clothed surface of the body, and so it is unnecessary to include the air speed in the assessment. Humidity has little effect on heat transfer unless the skin is wet with sweat, a condition usually associated with heat stress rather than thermal comfort. So for many indoor environments air temperature is a sufficient index.

MEAN WARMTH RESPONSES

In 1975 it was found that the mean warmth sensation of a group of people was usually close to thermal neutrality and marginally warmer than neutral (Figure 3.3). Two groups of respondents were exceptions. The first was a sample of Swedish schoolteachers surveyed on a spring day chosen for its high solar radiation, and the second was a sample of observations from people surveyed during the heat of the day in their flats in summer in the south of the former USSR. The conditions at the time of survey were not typical of their experience for either group, the typical being taken as the average room temperature during their waking hours over several preceding days. It was surmised that people felt hot if the indoor conditions were hotter than those to which they were currently adapted, rather than because of the conditions themselves (Nicol and Humphreys, 1973). It seemed that populations worldwide were quite well adapted to be comfortable in their

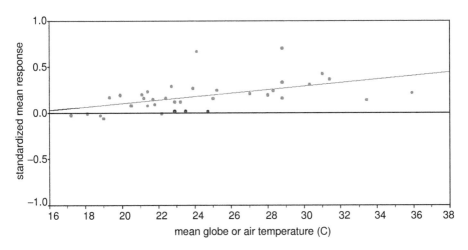

Note: Each point represents a separate batch of survey data. The mean responses were standardized to allow for the varied scales then in use, a value of unity corresponding to the extreme on the scale.
Source: Based on Humphreys (1975)

FIGURE 3.3 Scatter plot of standardized mean response and mean room temperature

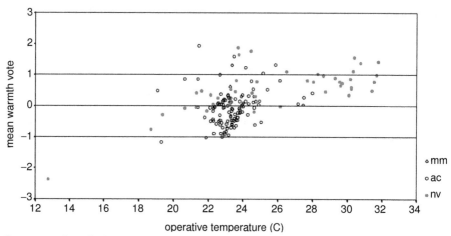

Note: nv=naturally ventilated, ac=centrally air-conditioned, mm=mixed mode operation.
Source: Humphreys et al (2005)

FIGURE 3.4 Scatter plot of mean warmth vote against mean operative temperature for the buildings in the de Dear database

own thermal environments, whether the temperature was as low as 17°C for elderly people in a UK winter, or as high as 34°C for Baghdad office workers in summer. Within this wide range the mean temperature had only a slight effect on the mean warmth reported by the respondents.

Do the subsequent data confirm this pattern? The de Dear database has results from 160 buildings and from each of these the mean warmth and the mean operative temperature can be calculated. Figure 3.4 shows these observations. A number of the mean responses lie quite far from thermal neutrality, and these occurred not only when the mean operative temperatures were exceptionally hot or cold. This suggests either that the adaptive response of these groups of occupants had failed fully to operate, or that the mean operative temperatures were not representative of the occupants' thermal experience. Unfortunately it is not possible to know how typical the temperatures were, but in 84 buildings in the database the survey lasted five or fewer days, and in 36 buildings the survey lasted just one or two days. The mean temperature for such short series can differ considerably from the mean over a longer series of days, so it is likely that it would sometimes differ from the longer-term average to which the occupants might be expected to have adapted themselves. The problem can to some extent be circumvented by pooling the data from several buildings for the same season and location, since the researchers often took measurements in several buildings in succession. The mean temperature from the pooled set of data would therefore be more typical of people's experience of the thermal environment.

Each point in Figure 3.5 represents a complete *file* from the database. (A file might be, for example, the data from several naturally ventilated buildings in Oxford in August. Air-conditioned buildings and naturally ventilated buildings are in separate files.) To further

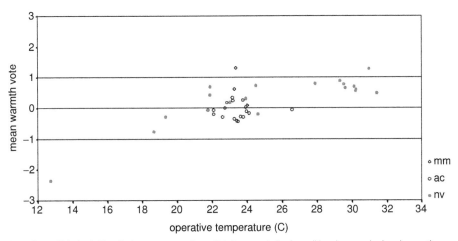

Note: N>=100 in the de Dear database, nv=naturally ventilated, ac=centrally air-conditioned, mm=mixed mode operation
Source: Humphreys et al (2005)

FIGURE 3.5 Scatter plot of mean warmth vote against mean operative temperature for the files

reduce the chance of atypical conditions, only files with 100 or more observations have been included. The figure shows that just three of the surveys had a mean ASHRAE vote falling outside the ±1 band. A mean indoor temperature of 12°C was found cool-to-cold (Saidu Sherif, North-West Frontier Province, Pakistan, January) while Bangkok office workers found 31°C slightly warm. Why a sample of office workers in northern England found 23°C similarly warm is unknown.

Figure 3.6 rests on the SCATs database. Each point represents a monthly survey of people in office buildings in a town in Europe, the data for the different heating modes being kept separate. Only three of the observations lie outside the ±1 band, and these are from surveys having very few observations.

Seasonal variation

It had been found in 1975 that, where the same population had been sampled in different seasons, people had adapted adequately to their seasonal indoor temperatures. These seasonal differences in mean indoor temperature were less than 6K. Absent from the data were populations experiencing larger seasonal differences. Recent data from five cities in Pakistan give us this extra information, for the seasonal variation of indoor temperature experienced in some of the cities was large. Figure 3.7 shows the mean comfort vote and the corresponding mean operative temperature for each monthly survey. It seems that people adapted moderately successfully to seasonal variation of monthly temperature up to about 10K, but were less successful outside this range.

The 1975 observation that the mean comfort vote from a survey was close to thermal neutrality, and showed but slight dependence on the mean indoor temperature during the survey, is therefore on the whole confirmed by the newer data. However, for this to apply

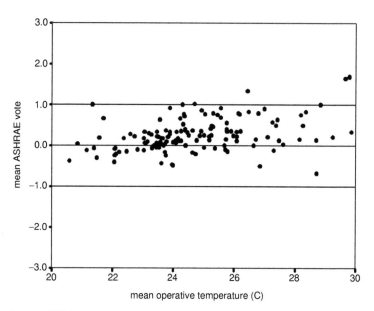

Source: Humphreys et al (2005)

FIGURE 3.6 Scatter plot of mean warmth vote against mean operative temperature for the SCATs database

it is necessary for the survey to represent the average of the temperatures experienced by the respondents over a substantial period. In the older surveys this was usually achieved inadvertently, because the laborious process of data acquisition ensured that the survey was likely to last a month or more. Recent studies commonly used automatic

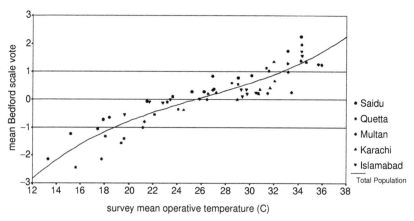

Note: The regression line is a cubic fit.
Source: Humphreys et al (2005)

FIGURE 3.7 Scatter plot of mean vote against mean operative temperature for data from monthly surveys in various cities in Pakistan

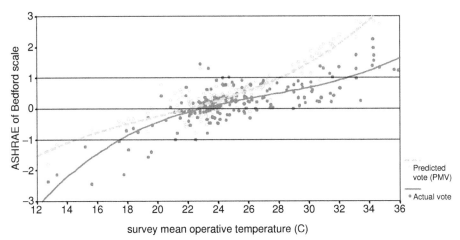

Note: Also shown are the corresponding values for PMV. The lines are cubic fits.
Source: Humphreys et al (2005)

FIGURE 3.8 Scatter plot of mean vote against mean operative temperature for all surveys listed in Table 3.1

data logging and completed the survey of a building in a day or two, thus increasing the likelihood of the mean operative temperature being unrepresentative of the longer-term experience of the occupants. The upward trend of the mean warmth sensation with rising temperature, although slight, is greater than had been found in 1975, and is in the region of one unit of the ASHRAE scale per 10K in both the de Dear and the SCATs databases. The recent data from Pakistan reveal that there are limits to the range of seasonal changes in indoor temperature for easy adaptation. Adaptation to seasonal change was substantially effective for seasonal changes up to about 10K. Changes greater than this may require customs or strategies that populations are unwilling to adopt or unable to use.

Figure 3.8 brings together the data from the de Dear database, the SCATs database and the other surveys (see Table 3.1). The figure shows the mean votes given by the various groups of respondents, together with the predicted values using the PMV equation. The fitted curves show that PMV deviates substantially from the respondents' mean thermal sensations at indoor temperatures higher or lower than those customary in temperate climates. The discrepancy between the sensation of warmth the people actually experienced and that predicted using PMV shows the limitation of using PMV to formulate thermal comfort standards for buildings, particularly for populations in warm accommodation, where PMV quite seriously overestimates the subjective warmth. This difference could be important for energy consumption and for the design decisions of the architect and engineer. PMV was at its best when the mean indoor temperature was in the region 21–25°C.

MEAN CLOTHING INSULATION

The chief means people use to adapt to the differing temperatures in their accommodation is by choosing clothing of suitable thermal insulation. Surveys collected

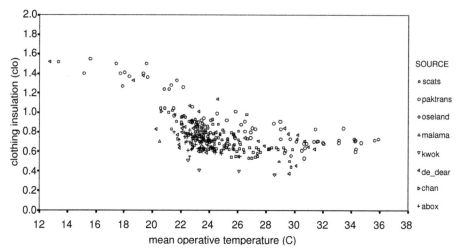

Note: Each observation is the joint-mean of a separate survey of thermal comfort.
Source: chapter authors

FIGURE 3.9 Mean clothing insulation and mean room temperature

in 1975 had only general descriptions of the clothing and few gave estimates of its thermal insulation. Recent surveys usually have estimates of clothing insulation from tables provided in ISO 7730 or ASHRAE Standard 55. Figure 3.9 is a scatter diagram of the mean clothing insulation against the mean operative temperature in the accommodation.

A dependence on temperature is evident. The greatest rate of change of clothing with temperature occurred in the range 18–28°C. Below 18°C people did not seem to wear more clothing in colder environments, while above 28°C people did not seem to wear less clothing in warmer environments. There were notable differences between the clothing of different groups at the same temperature. In particular Kwok's students in Hawaii were lightly clad (~0.4clo) compared with Pakistani office workers at similar temperatures (~0.9clo). This difference could be attributable to social and cultural influences to which clothing is subject. In Pakistan it was not acceptable to have the limbs uncovered, while in Hawaii there was no such constraint. It may also be that the insulation of the traditional summer lightweight shalwar-kameez worn in Pakistan is overestimated, and results from thermal manikin measurements are awaited. The values for the mean clothing insulation ranged from 0.4–1.6clo-units. This would be sufficient to offset a temperature difference of some 7K, much less than the range of the mean operative temperatures to which people adapt (see below). So it is evident that other means of adaptation are also being employed, if we may assume that the estimates of clothing insulation are not grossly misleading.

Other controls

People use other means to make themselves thermally comfortable, such as adjusting windows, blinds, fans and the heating and cooling system. The use of such controls is evidence of adaptation and can affect the energy used by the building. The ability of

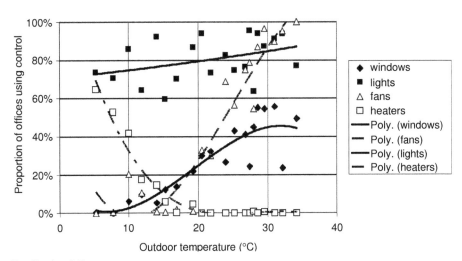

Note: Data from Pakistan.
Source: Nicol and Humphreys (2004)

FIGURE 3.10 Probability of use of various building controls: Variation with outdoor temperature

building occupants to control their environment has been referred to as the 'adaptive opportunity' afforded by the building (Baker and Standeven, 1995). Increased adaptive opportunity reduces discomfort and dissatisfaction (Leaman and Bordass, 1997) and this finding is entirely consistent with the adaptive approach. There may, however, be conflicts, for instance between the cooling effect of opening a window and the resulting increased noise from outside.

A full account of occupants' use of controls is beyond the scope of this review, but an example from the Pakistan database is given. Figure 3.10 illustrates how the use of various controls changes with the outdoor temperature. Controls whose effect could be to cool the occupants (fans, windows) are more likely to be used as the outdoor temperature rises, whereas heaters are less likely to be used. The use of lights hardly changes with temperature, depending rather on the external illuminance (Sutter et al, 2006). A recent analysis by Nicol and Humphreys (2004) suggested that the use of controls was consistent with the need to change the thermal environment to suit occupants and that the likelihood of a particular control being used could be quantified using probit analysis. Quantified data on the use of controls is needed to inform thermal simulations of occupied buildings to enable them properly to evaluate the effect of occupant behaviour on comfort and energy use (Bourgeois et al, 2006), and this is an area of current research.

REGRESSION GRADIENTS

Regression analysis may be used to estimate the variation of warmth vote within a survey in response to the variation of temperature. It produces an equation relating the expected value of the thermal sensation to the room temperature:

$$TSV = a + b \times T_{op} \tag{1}$$

where TSV is the thermal sensation vote on a seven-point warmth scale such as the ASHRAE scale, T_{op} the corresponding operative temperature, a the intercept on the sensation axis, and b the regression coefficient. The procedure is illustrated in Figure 3.11. The regression coefficient indicates how much the comfort vote increases per degree increase in operative temperature, and is a measure of sensitivity to temperature change. In the 1975 database there were 11 surveys that had used a seven-category scale and from which a value could be extracted. The mean regression coefficient (for surveys of typically a month or so duration) was 0.24 scale units/K, lower than the 0.32/K that had been found in climate chamber studies (McIntyre, 1980). This was not surprising because in daily life people adjust their clothing to alleviate to some extent the effects of changing temperatures. Surveys conducted over more lengthy periods yielded lower regression coefficients, suggesting that the longer the period the more complete was the adaptation.

The de Dear database and the SCATs database provide a large number of values for the regression coefficient of comfort vote upon operative temperature. From the de Dear

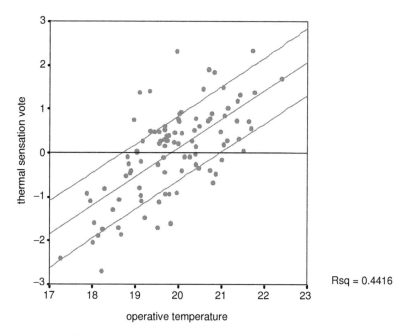

Note: A continuous form of the thermal sensation scale has been used for this illustration. The central line is the regression line; its gradient is the regression coefficient. The outer lines indicate the standard deviation of the scatter of the observations about the line in the vertical direction.
Source: chapter authors

FIGURE 3.11 Illustration of the regression procedure: Scatter plot of subjective warmth and operative temperature

database a regression coefficient can be extracted for each building during each survey (usually lasting a few days), while from the SCATs database a value can be extracted for each building during each monthly survey (usually lasting a day or two). Many of these estimates are of low precision, either because of the limited number of observations or because of the narrow range of operative temperature.

The estimated regression gradients are shown in Figure 3.12 (de Dear database, 143 regressions) and Figure 3.13 (SCATs database, 99 regressions), plotted against the standard deviation of the operative temperature during the period of each of the surveys. Regressions based on fewer than 20 observations have been discarded. It could not be said with confidence that the different modes of operation of the buildings affect the values of the regression coefficients, for the overlap among the modes is extensive. Statistical analysis shows that the regression gradient is significantly higher in the air-conditioned buildings ($p=0.026$) in the de Dear database, while there are no significant differences in the regression gradients among the modes in the SCATs database, but the tendency is in the opposite direction. Pooling the data from the two databases yields no significant difference between the regression gradients for the air-conditioned mode and other modes of operation.

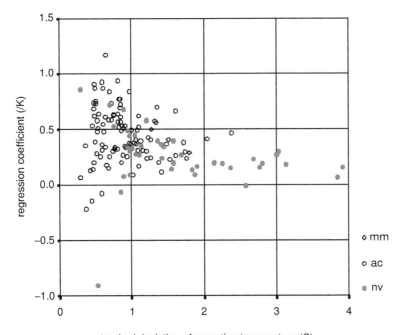

Note: de Dear database N>=20.
Source: Humphreys et al (2005)

FIGURE 3.12 Scatter plot of regression coefficients of warmth vote on operative temperature, against the standard deviation of the operative temperature (de Dear database)

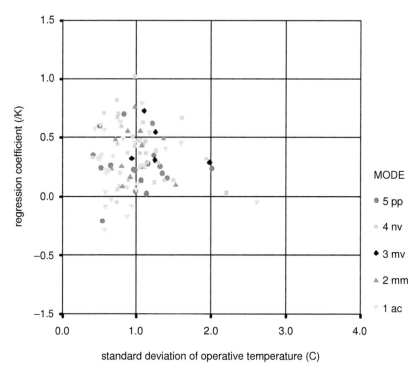

Note: SCATs database N>=20. pp=some rooms air-conditioned, mv=mechanical ventilation.
Source: Humphreys et al (2005)

FIGURE 3.13 Scatter plot of regression coefficients of warmth vote on operative temperature, against the standard deviation of the operative temperature (SCATs database)

In both sets of data the regression gradient appears to decrease as the standard deviation of the operative temperature increases. However, there are few data for the higher standard deviations, and the trend cannot be estimated with useful precision. Assuming the effect to be real, there are two kinds of explanation that could be offered:

● *Behavioural* – if the room temperature has a wider range it would be expected that people would change their clothing to compensate to some extent for the change in room temperature. There could be other changes too, also dependen on the room temperature, such as the use of fans or changes in posture (Raja and Nicol, 1997). All these effects together would depress the gradient of the regression line.
● *Psychological* – there can be a tendency, when subjective rating scales are used, for respondents to adjust their voting to accommodate the range of the subjective scale to the range of conditions they experience. The range effect would suggest that people accustomed to a wide range of room temperature might be less sensitive to change than those accustomed to smaller variations. The effect has

been seen in estimations of noise nuisance (see, for example, Poulton, 1973; 1975), but has not as far as we are aware been demonstrated for the thermal environment, where the sensation of warmth is largely dependent on the thermal state of the body.

There is also an effect associated with the assumptions underlying the statistical analysis. Regression analysis assumes that the predictor variable, in this case the operative temperature, is without error, and if error in the predictor is substantial the regression gradient is depressed. Two kinds of error may be present in the predictor:

● *Measurement error* – the operative temperature is not difficult to measure, so it is probable that its measurement error is quite small, and would depress the regression gradient only if the range of operative temperatures were also very small.
● *Equation error* – this error refers to the incompleteness or insufficiency of the predictor variable as an index to evaluate the dependent variable. The effect on the regression coefficient diminishes as the range of the predictor variable increases, unless the equation error expands as the range of the variable increases. The operative temperature is an incomplete predictor of environmental warmth, since it neglects variations in humidity, air movement, clothing and activity, so we might expect the regression gradient to suffer depression when the standard deviation of operative temperature is small. Unfortunately the size of the effect cannot be satisfactorily estimated for the lack of a perfect index with which to compare it. (For a thorough exposition of regression with errors in the predictors, see Cheng and van Ness (1999), and for an exploration of the effect in relation to the de Dear database, see Humphreys and Nicol (2000). It would at first be thought that the problem of the insufficiency of the operative temperature could be overcome by introducing into the regression further terms for air movement, humidity, clothing and metabolic rate. Unfortunately the error in the estimation of some of these terms nullifies the theoretical advantage of including them, in a manner similar to that noted earlier in the discussion of thermal indices.)

Estimating the regression coefficient

These complications make it difficult to estimate a regression gradient that indicates the ordinary sensitivity of the respondents to the modest room temperature changes normally encountered during the course of a few working days. Figure 3.14 summarizes the information from the two databases. The regression gradient has a maximum value when the standard deviation of the operative temperature is in the region of 1K, a value typical for occupied rooms during a brief survey period (if the temperature distribution were normal this standard deviation would imply that the temperature remained within a $\pm 2K$ band for 95 per cent of the time). Below 1K standard deviation of temperature the regression gradient is slightly depressed, probably because of the effects of measurement error and equation error, while above this value it decreases, as would be expected from the effect of increased behavioural adaptation and any psychological range effect. It is perhaps therefore best to take the maximum of the curve to indicate the ordinary value of the coefficient. This gives a regression gradient of 0.40 ± 0.02 scale units

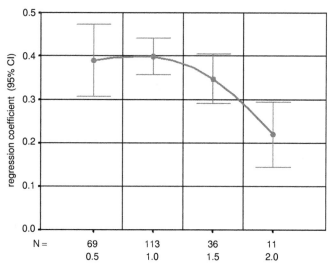

N =   69   113   36   11
        0.5   1.0   1.5   2.0

Standard deviation of operative temperature (C)

Note: The error-bars indicate the 95 per cent confidence intervals of the mean of the coefficients.
Source: Humphreys et al (2005)

FIGURE 3.14   Summary of mean values of the regression coefficients (/K), in relation to the standard deviation of the operative temperature, from the de Dear and the SCATs databases

per degree. The value is higher than the value of 0.33/K established by climate chamber research for sedentary people in standard clothing (0.6clo). The mean value from the field data is 0.37±0.02, considerably higher than the mean value of 0.24 scale units/K obtained in 1975. The difference is probably attributable to the older surveys being conducted over more protracted periods, and the coefficient therefore prone to dilution because of behavioural adaptation during the survey period.

INTERPERSONAL DIFFERENCES IN THERMAL SENSATION

The residual standard deviation is the measure of the scatter of the individual comfort votes around the regression line (see Figure 3.10). It represents the differences between individuals in their warmth response, or, if the same people vote more than once, the differences in response of the same people on various occasions. Most research designs have been cross-sectional. That is to say, during any fairly brief survey each person responds on just one occasion, although the same respondents may be revisited during subsequent surveys. So in each survey the residual standard deviation measures the inter-subject variation. In the 1975 analysis this standard deviation of individual differences was derived from probit analysis (Finney, 1964). We have not used probit analysis, but if the assumptions underlying regression analysis and probit analysis are fulfilled, the two methods, although apparently so different, are formally equivalent and therefore yield similar results. Table 3.4 shows the residual standard deviations from three databases. It

TABLE 3.4 Residual standard deviations after regression

DATABASE	MODE	NO. OF SURVEYS	RESIDUAL STANDARD DEVIATION
De Dear	a.c	111	1.09
	n.v	44	0.94
	m.m	4	1.09
SCATs	Various	178	1.10
Abdnox	a.c.	54	1.06
	n.v.	124	1.12
		Mean value:	1.07

Note: a.c.=air-conditioned, n.v.=natural ventilation, m.m.=mixed mode.

is remarkably constant across the databases, irrespective of the mode of heating or cooling. The mean is 1.07 scale units, and corresponds to 2.7K if the regression gradient is 0.4/K. The corresponding values found in 1975 were 2.8K for non air-conditioned rooms, and 1.8K for the few air-conditioned spaces in the sample.

DISTRIBUTION OF THERMAL COMFORT

In 1975 charts were presented to summarize the effects on comfort of departures from the optimum temperature for the group. Separate curves were derived for adults in air-conditioned (AC) and non air-conditioned accommodation, and for young children at school. The curves for adults had as their basis an interval of 4K between the scale categories, and a standard deviation of individual differences of 3K in non-AC and 2K in AC accommodation. Analysis of the more recent data did not confirm a systematic difference between the two modes of operation in either the sensitivity to operative temperature or the extent of individual differences. A single chart for adults is therefore given (Figure 3.15), applicable to either mode. It assumes a regression gradient of 0.4/K, equivalent to an interval of 2.5K between the ASHRAE scale categories, and a standard deviation of individual differences of one scale unit, also equivalent to 2.5K. There are no new data that we are aware of for young children. The upper line on the chart uses the conventional assumption that votes in the central three categories of the ASHRAE scale correspond to comfort.

Implications for thermal comfort standards

The maxima on charts of the type shown in Figure 3.15 are sensitive to even quite small changes in either the regression gradient or the standard deviation of the individual differences, and hence the estimate of the proportion of people in discomfort is prone to error. In current standards for thermal comfort (for example, ASHRAE, 2004, ISO 7730, 2005) the limits on operative temperature or of PMV are expressed in terms of a percentage of people expected to be uncomfortable – typically set at 20 per cent. These predictions of discomfort are unlikely to bear much relation to its actual extent. The same difficulty is inherent in the curve of PPD against PMV. The predicted discomfort in everyday conditions bears little relation to its actual prevalence (de Dear et al, 1997; Humphreys and Nicol, 2002).

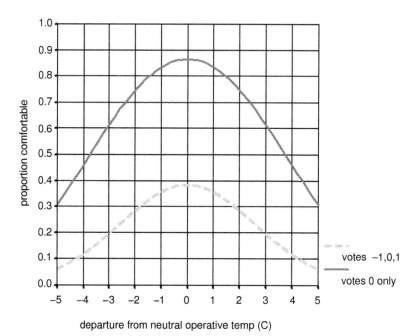

Note: The figure is based on a regression gradient of 0.4/K and a standard deviation of 2.5K.
Source: Humphreys et al (2005)

FIGURE 3.15 Proportion of people likely to be comfortable

TEMPERATURES FOR THERMAL NEUTRALITY

If the mean thermal sensation vote ($TSV_{(mean)}$) and mean operative temperature ($T_{op(mean)}$) are known, and a reliable value of the regression gradient (b) is supplied, it is easy to calculate the operative temperature ($T_{op(n)}$) at which the respondents would have on average reported being thermally neutral:

$$T_{op(n)} = T_{op(mean)} - TSV_{(mean)}/b \qquad (2)$$

From the previous section it is apparent that it is advisable to use a standard regression gradient, rather than the empirical value from each individual survey, because the estimate of the latter is prone to much uncertainty. An appropriate value is 0.40/K, as shown above. This will give the operative temperature at which the average sensation would be zero ('neutral' on the ASHRAE scale, 'comfortable' on the Bedford scale). A sufficiently precise estimate of the neutral temperature can be made even if the survey has rather few comfort votes, which may be shown as follows: the residual standard deviation (s.d.) of a set of comfort votes on either the ASHRAE or the Bedford scale is approximately one scale unit, so the standard deviation of the mean (s.d.m.) of a group of 25 votes would be some 0.2 units (s.d.m. = s.d./√n). This corresponds, if the regression gradient is 0.4/K, to a standard error of 0.5K for the estimate of the operative temperature for thermal neutrality for the group.

Applying this method to the available surveys, the temperatures for neutral on the ASHRAE scale, or for comfort on the Bedford scale, have been estimated. They range from 18°C for Saidu Sherif, North-West Frontier Province, Pakistan in winter, to 33°C for Multan, Pakistan in summer. The air movement in Multan averaged 0.7m/s, so if the air movement had been slight, the neutral temperature would have been about 31°C. This wide range of neutral temperatures is similar to that found in 1975, which ranged from 17°C for elderly people at home in winter in the UK to 33°C for office workers in Baghdad, Iraq in summer, which also would have been equivalent to about 31°C had the air movement been slight (Nicol, 1974). Now, as then, the difference between the neutral temperatures of the two populations is far greater than can be explained by the difference in clothing insulation alone (Saidu: 1.5clo, Multan: 0.7clo), and it seems necessary to conclude that the two populations describe as 'neutral' or 'comfortable' different body-states.

Figure 3.16 is a scatter diagram of the neutral temperatures against the mean of the operative temperature at the times when their comfort votes were obtained. Each of the points represents an independent survey, with surveys having fewer than 20 observations being discarded. There is a strong dependence of the neutral temperature on the mean

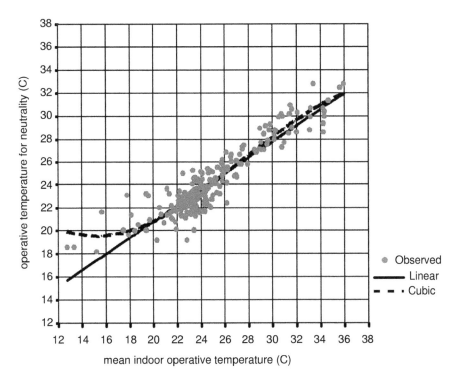

Source: Humphreys et al (2005)

FIGURE 3.16 Scatter plot of neutral temperatures against the mean operative temperature

operative temperature experienced during the survey. The equation, if a linear relation is assumed, is:

$$T_n = 6.89 + 0.70T_{op(mean)} \qquad r = 0.92 \qquad (3)$$

where r is the Pearson Correlation Coefficient.

The equivalent relation found in 1975 was:

$$T_n = 2.56 + 0.83T_{op(mean)} \qquad r = 0.96 \qquad (4)$$

The new data show a slightly lower dependence on the mean temperature, but their range includes mean indoor temperatures as low as 13°C, compared with a low of 17°C in the earlier data. It seems that for the lowest mean temperatures, and perhaps for the highest, there is a tendency for the relation to 'flatten' as people's adaptive range approaches its practical limit. A curvilinear fit could therefore be appropriate, and a third order fit is shown in Figure 3.16. The curve suggests that adaptation had become less effective when the room temperatures were persistently much below 20°C. The new data thus confirm the strength of the tendency observed in 1975 for different populations to be adapted to very different indoor temperatures, but they also begin to indicate the limit of the range of easy adaptation for some populations.

DO PEOPLE LIKE TO FEEL 'NEUTRAL'?

The question was raised in 1975 whether the centre point on the warmth scales necessarily corresponded to the optimum for comfort, and it was suggested that perhaps people in warm climates liked to feel cooler than neutral, while people in cold climates might prefer to feel warmer than neutral. In many of the more recent thermal comfort surveys this question has been addressed by supplementing the ASHRAE or Bedford scale with a separate preference scale. Preference scales enquire whether the respondent would like to feel warmer or cooler, or whether they desire no change (see Table 3.2). Whereas the use of a seven-category scale has become virtually standard for assessing subjective warmth, the number of categories for thermal preference scales varies from three to seven. When field surveys have been analysed, differences between the 'preferred temperature' and the 'neutral temperature' have frequently been found, and the difference may be as much as three degrees, but generally it is much smaller.

An analysis of the behaviour of the ASHRAE scale in the de Dear database and in the SCATs database (Humphreys and Nicol, 2004) found that the people did prefer to feel cooler when the outdoor temperature was higher, but also that they preferred to feel warmer when the *indoor* temperature was higher (Figure 3.17). There was much variation between surveys, and as with regression coefficients, the value obtained from any one survey is subject to substantial error. The systematic shift of the preferred point on the ASHRAE scale is 0.33 scale units or less, which, for a regression gradient of 0.4/K, corresponds to a shift in neutral temperature of 0.8K or less. The effect is likely to be smaller for the Bedford scale, where the centre of the scale is labelled 'comfortable' rather than 'neutral'. Figure 3.16 would therefore be little affected by including an allowance for a disparity between 'neutrality' and 'preference'.

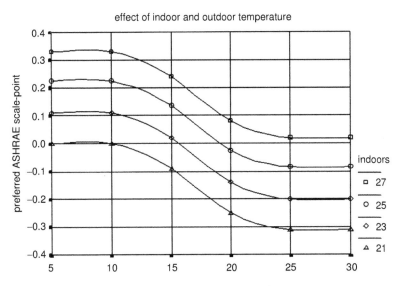

effect of indoor and outdoor temperature

Source: Humphreys and Nicol (2004)

FIGURE 3.17 Chart showing the effects of indoor and outdoor temperature on the preferred point on the ASHRAE scale

Note on the McIntyre scale

It has perhaps too readily been assumed that the McIntyre scale ('prefer warmer', 'prefer cooler', or 'prefer no change') is unambiguous in its definition of comfort. Heidari, from wide experience of surveys in Iran reports that many who say they are perfectly comfortable also say they would 'prefer cooler'. The response to the McIntyre scale seems in this context to reveal a lifestyle aspiration rather than a comfort state: given the choice some people would prefer to wear heavier clothing and be in a cooler room (Heidari, S. personal communication, 2006).

Note on the ASHRAE scale

Recent work has shown that people differ systematically from each other in the position on the ASHRAE scale that they consider most desirable, and that this position itself depends on how warm or cool the person is currently feeling (Humphreys and Hancock, 2007). It has also been shown that the intervals on the scale differ according to the language in which it is presented (Pitts, 2006). The behaviour of the scale is therefore more complex than has generally been assumed.

OUTDOOR TEMPERATURES AND COMFORT INDOORS

The indoor comfortable or neutral temperature is affected by the prevailing outdoor temperature in a manner that depends on the mode of operation of the building (Humphreys, 1978):

- *Free-running mode.* In the free-running mode, where no energy is being supplied for heating or cooling the accommodation, the indoor temperature will drift under the influence of the day-to-day changes and the seasonal variations of the outdoor temperature. The indoor temperature is moderated by the thermal capacity of the building, and can to an extent be controlled by opening or closing windows, and by using blinds to control penetration of solar radiation. Because the occupants tend to adjust to their prevailing indoor temperature, chiefly by the choice of appropriate clothing, the comfort temperature also drifts in concert with the outdoor temperature. Thus there arises a statistical relation between the temperature for comfort indoors and the prevailing outdoor temperature. It is governed by a combination of building physics and human behaviour. A precise relation is not to be expected, because buildings of different climatic design provide different indoor temperatures, to which the occupants will tend to adapt.
- *Heated or cooled mode.* The free-running mode ceases to function satisfactorily if the building is inconveniently cold or hot for the occupants, so heating or cooling is then desirable. The heated or cooled mode typically operates in naturally ventilated buildings during the heating season and in centrally air-conditioned buildings all year round. The dependence of the comfort temperature on the prevailing outdoor temperature is weaker than for the free-running mode. Nevertheless, people dress according to weather and season, and this variation of clothing insulation is reflected in a corresponding variation in the temperatures that are desired indoors. Also, when the outdoor temperatures are pleasant it may be desirable to reduce the disparity between indoors and outdoors to avoid discomfort on entering or leaving the building. Therefore some dependence of indoor comfort temperature on the prevailing outdoor temperature would still occur. When outdoor

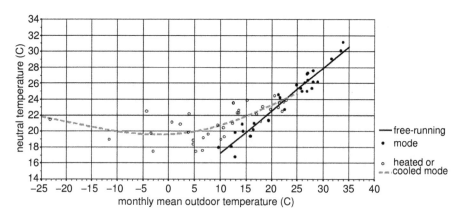

Note: The trend curves are fitted by Lowess regression (50 per cent, three iterations).
Source: Based on Humphreys (1978)

FIGURE 3.18 Scatter plot for neutral temperatures for the 1978 data

temperatures are more extreme, the indoors becomes a haven from the rigours of the climate, and the temperature for comfort indoors will bear little if any relation to outdoor conditions. It becomes a matter of social custom, different temperatures being characteristic of different societies both historically and geographically (Shove, 2003).

Figure 3.18 shows the data published in 1978. For the free-running mode there was a surprisingly definite linear relation between the monthly mean outdoor temperature and the indoor neutral temperatures (r=0.97). For the heated and cooled mode there was a weaker and curvilinear relation (r=0.72). There were no data from air-conditioned buildings in hot climates. If the two modes were pooled, the resultant curvilinear relation had an intermediate correlation (r=0.88).

Changes since 1978

The distinction between the two types of operation is less clear now than it was 30 years ago. Oil- and gas-fired heating systems can be operated on demand from thermostats, and the 'heating season' typical of buildings with coal-fired boilers is largely a thing of the past. Mixed mode buildings have also been introduced. They provide winter heating and summer cooling when needed, but have a zone of drifting temperature between the heating and the cooling mode. So it is sometimes not clear whether, during a survey, a building was operating in the free-running or the heated or cooled mode. This is particularly true when the outdoor conditions are such that heating might or might not be required. Another change has occurred during the intervening years: there has in many countries been a progressive increase in the level of thermal insulation of buildings, together with an increase in the use of

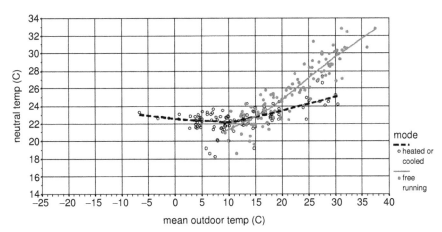

Note: The trend curves are fitted by Lowess regression (50 per cent, three iterations). Surveys having fewer than 20 votes have been discarded. Surveys where the mode was intermediate have been omitted.
Source: Humphreys et al (2005)

FIGURE 3.19 The dependence of the neutral temperature on the prevailing outdoor mean temperature for the free-running and the heated/cooled modes of operation

electrical equipment such as computers, printers and copiers. The combined effect has been to alter the relation between indoor and outdoor temperature for a building in the free-running mode. A consequent effect on the comfort temperatures is to be expected.

The recent data

Figure 3.19 shows the equivalent figure using the post-1978 data. The general pattern is the same. For the free-running mode there remains a strong relation, and the curve of the heated or cooled mode is replicated. However, the neutral temperatures are higher by about 2K than they were in 1978. This reflects the fact that the indoor temperatures were in general some 2K higher than they were in the older sample, and the occupants are adapted to these higher temperatures. That buildings in general run at higher temperatures than they did some 30 years ago is probable, but it is not a necessary inference from these data, for in neither the older nor the more recent surveys did researchers select representative samples of occupied buildings.

The use of a regression line to represent each mode of operation has led to misunderstanding. The scatter of the points around the line is not merely error, since the standard error of each point (as shown above) is 0.5K or less. Most of the scatter therefore represents genuine differences of comfort temperature between different groups at the same outdoor temperature. It follows that deviations from the line should not be used to estimate probable percentages of people in discomfort, since this depends on the distance from the relevant point rather than from the line.

Representing the outdoor temperature

It is not obvious how best to represent the outdoor temperature when using it as a base from which to indicate probable indoor comfortable temperatures. In 1978, in the absence of contemporary data, the historic monthly maximum and minimum temperatures from meteorological stations were used. The neutral temperature correlated far better with the simple average of the maximum and minimum temperatures than with either the maximum or the minimum alone, and the correlation could not be improved by adjusting the relative weights allocated to the maximum and minimum temperatures. The relative success of this simple average is probably attributable to the pervasive effect of the average outdoor temperature on the temperatures in the buildings and on the clothing of the people. However, it is likely that improved measures can be developed, for two main reasons:

- *Speed of human response.* Successful adaptation to changing temperatures is achieved, to a large extent but not exclusively, by changing the clothing according to the prevailing conditions. Change of clothing in response to indoor temperature is likely to be small within any particular working day, but becomes effective over a period of a few days (see, for example, Humphreys, 1979). A month is therefore likely to be too long a period over which to average the outdoor temperature.
- *Year-to-year variation.* The mean outdoor temperature in a particular month, in some climates, varies considerably from year to year. It therefore seems fitting to use the

actual prevailing temperature rather that the 20 or 30 year average provided by meteorological tables.

An exponentially weighted running mean of the daily average is an attractive representation of the outdoor temperature because it gives greatest weight to recent experience and progressively reduced weight to the temperature on earlier days. It is simple to calculate, since it can be updated daily. The new running mean ($T_{rm(new\ day)}$) is the weighted average of the previous value ($T_{rm(previous\ day)}$) and the mean temperature of the new day in the sequence ($T_{mean(new\ day)}$). It seems that the best weighting is about 0.8 of the running mean and 0.2 of the new day's mean (see, for example, McCartney and Nicol, 2001; Nicol and Humphreys, 2002), thus:

$$T_{rm(new\ day)} = 0.8T_{rm(previous\ day)} + 0.2T_{mean(new\ day)} \qquad (5)$$

Conclusions on the relation between indoor comfort and outdoor temperature

The basic pattern of the relation between outdoor temperature and comfort indoors, first described in 1978, is evident also in the newer data. However, the temperatures for comfort appear to have risen over the years, both for free-running and in heated and cooled modes of operation. This shift is related to the generally higher indoor temperatures in the more recent data. The use of the tabulated historic monthly outdoor temperature is being replaced by more responsive measures, such as the exponentially weighted running mean.

THE ADAPTIVE APPROACH AND STANDARDS FOR THERMAL COMFORT

By its nature the adaptive approach does not lend itself to proposing fixed temperatures for occupied buildings; people continually adapt to a variety of temperatures, as the field data amply demonstrate. So a truly adaptive standard would be 'shaped' differently from present standards. It would aim to promote easy adaptation, and would include the following parts:

● *Customary temperatures.* Populations are most likely to be adapted to the temperatures they normally experience. There are two reasons for this: indoor temperatures tend to be adjusted to meet the requirements of the occupants, and occupants tend to adapt their clothing and customs to their indoor temperatures. In this way there arise in societies conventional or usual temperatures that the occupants find to be comfortable. This process of convergence will operate satisfactorily unless the constraints posed upon it are excessive. Constraints may arise, for example, from the prohibitive cost or unavailability of fuel for heating and cooling, or from inhospitable climates, or from social pressures on dress that limit people's choices. If the constraint is less severe, it will not prevent convergence from being achieved, but will alter the temperature at which it occurs. Thus, for example, the temperatures customary for comfort vary geographically and seasonally with the climate, and may be modified by changing affluence. So the most direct way to set up standards for thermal comfort would be to ascertain, for the relevant society, what temperatures are customary in buildings of different kinds (offices, schools, theatres, homes) at different seasons of the year, and to use these temperatures as the basis of the standards. (It is surprising that in most countries these customary temperatures are not established, and temperature measurement in representative samples of buildings is rarely undertaken. An exception is New Zealand, where such a programme is in place for temperatures in dwellings (see, for example, BRANZ, 2005).)

● *Adaptive opportunity.* A feature of an adaptive standard would be to ensure that the building afforded sufficient 'adaptive opportunity' so that convergence could operate unimpeded. This would be achieved by the provision of effective and user-friendly control over the thermal environment, whether by temperature controls, openable windows, solar controls or the provision of controllable fans. An adaptive standard would indicate what range of control might be required.

● *Flexible dress.* A further feature would be recommendations about flexibility in dress codes, since the choice of suitable clothing is important for achieving comfort.

● *Thermal stability.* The process of adaptation is easier if the environment in the building is not prone to sudden unpredictable changes, since it is difficult to adapt to rapidly changing conditions. An adaptive standard would indicate what degree of stability to provide.

The adaptive approach is beginning to influence standards and guidelines for comfort in buildings, as is evident from the use of field results in ASHRAE Standard 55-2004 (ASHRAE, 2004) and (in the UK) the Chartered Institute of Building Services Engineers Guide (CIBSE, 2006). An adaptive section is also included in the new European standard EN 15251 (CEN, 2007).

CONCLUSION

This paper compared the field study thermal comfort data available in 1975–1978 with data subsequently collected. It has found that the patterns evident from analysis of the earlier data are also evident in the more recent data, although there are a number of interesting changes.

First, multi-term indices of thermal comfort continue to disappoint in their ability to predict the thermal sensation of the occupants. The rational indices that take into account the personal variables of clothing and metabolic rate (PMV, SET) are inferior to the air temperature or the operative temperature in their correlation with people's sensation of warmth.

Second, populations continue to establish thermal comfort in a diversity of indoor temperatures, the range being greater than predicted by the physiological models of thermal comfort. It follows that these models do not sufficiently allow for human adaptability to the thermal environment. However, recent data from year-round studies show that if the seasonal differences in indoor temperature exceed about 10K, it is likely that the some occupants will not adapt sufficiently and may experience discomfort during the seasonal extremes.

Third, the sensitivity of people to within-day and day-to-day changes of temperature is greater than expected from the earlier data. The regression gradient of the ASHRAE or Bedford scale vote on the operative temperature is approximately 0.40 scale units per degree, compared with some 0.24/K in the older data. This is probably not a change in the behaviour of the people or the scales, but a feature disclosed by the rapidity of modern data collection. This has enabled data to be collected in a few days that would typically have taken a few weeks 30 years ago. The newer data thus eliminate from the regression behavioural adaptation that occurs over the longer period.

Fourth, there remains a strong correlation between the mean room temperature during a survey and that found to be thermally neutral or comfortable by the occupants. It is, however, necessary for the room temperature to be averaged over a sufficient period of time, so that it can be expected to represent the usual condition the people experience in their accommodation. The newer data reveal that at room temperatures much below 20°C some people may experience persistent cold discomfort.

Fifth, the pattern of the relation between the prevailing mean outdoor temperature and the temperatures indoors for comfort or neutrality remains unchanged in form, but the comfort temperatures are higher by some 2K, both in free-running and heated/cooled modes of operation. This reflects generally higher indoor temperatures in the recent data, and would be an expected adaptive response. It is likely that indoor temperatures have risen over the intervening years, but a rise cannot be demonstrated because neither the earlier nor the later data are from representative samples of buildings.

AUTHOR CONTACT DETAILS
Michael A. Humphreys, School of Architecture, Oxford Brookes University, Gipsy Lane Campus, Headington, Oxford, OX3 0BP, UK
J. Fergus Nicol, School of Architecture, Oxford Brookes University, Gipsy Lane Campus, Headington, Oxford, OX3 0BP, UK
Iftikhar A. Raja, COMSTATS Institute of Information Technology, University Road, Abbottabad, Pakistan

ACKNOWLEDGEMENTS
An earlier version of this review was given at the Comstats Institute of Technology, Abbottabad, Pakistan, in June 2005 (Humphreys et al, 2005).

REFERENCES
ASHRAE (2004) *ANSI/ASHRAE Standard 55-2004: Thermal Environmental Conditions for Human Occupancy,* American Society of Heating, Atlanta, Georgia, Refrigerating and Air-conditioning Engineers (ASHRAE)

Auliciems, A. (1981) 'Towards a psycho-physical model of thermal perception', *International Journal of Biometeorology*, vol 25, pp109–122

Baker, N. V. and Standeven, M. A. (1995) 'A behavioural approach to thermal comfort assessment in naturally ventilated buildings', *Proceedings CIBSE National Conference*, Chartered Institute of Building Service Engineers, London

Bedford, T. (1936) *The Warmth Factor in Comfort at Work*, Medical Research Council Industrial Health Board, Report No 76, HMSO, London

Bourgeois, D., Reinhart, C. and Macdonald, I. (2006) 'Adding advanced behavioural models in whole building energy simulation: a study on the total energy impact of manual and automated lighting control', *Energy and Buildings*, vol 38, no 7, pp814–823

BRANZ (2005) *Energy use in New Zealand Households*, Study Report No. SR 133, BRANZ, New Zealand

CEN (2007) *CEN 1521: Indoor Environmental Input Parameters for Design and Assessment of Energy Performance of Buildings Addressing Indoor Air Quality Thermal Environment, Lighting and Acoustics*, European Committee for Standardization, Brussels

Chan, D. W. T., Burnett, J., de Dear, R. J. and Ng, S. C. H. (1998) 'A large scale survey of thermal comfort in office premises in Hong Kong', *ASHRAE Technical Data Bulletin*, vol 14, no 1, pp76–84

Cheng, C.-L. and van Ness, J. W. (1999) *Statistical Regression with Measurement Error*, Kendall's Library of Statistics 6, Arnold, London

CIBSE (2006) *The Guide, Section A1, Comfort*, Chartered Institute of Building Service Engineers, London

de Dear, R. J. (1998) 'A global database of thermal comfort experiments', *ASHRAE Technical Data Bulletin*, vol 14, no 1, pp15–26

de Dear, R. J. and Brager, G. S. (1998) 'Developing an adaptive model of thermal comfort and preference', *ASHRAE Technical Data Bulletin*, vol 14, no 1, pp27–49

de Dear, R. J., Brager, G. and Cooper, D. (1997) *Developing an Adaptive Model of Thermal Comfort and Preference*, Final Report on RP 884, Macquarie University, Sydney

Fanger, P. O. (1970) *Thermal Comfort*, Danish Technical Press, Copenhagen

Fiala, D. (1998) *Dynamische Simulation des menschlichen Warmehaushalts und der thermischen Behaglichkeit*, PhD thesis, De Montfort University and the Fachhochschule Stuttgart, Hochdchule fur Technik, Leicester and Stuttgart

Fiala D., Lomas, K. J. and Stohrer, M. (1999) 'A computer model of human thermoregulation for a wide range of environmental conditions: The passive system', *Journal of Applied Physiology (American Physiological Society)*, vol 87, no 5, pp1957–1972

Finney, D. J. (1964) *Probit Analysis: A Statistical Treatment of the Sigmoid Response Curve*, Cambridge University Press, Cambridge

Gagge, A. P., Fobolets, A. P. and Berglund, L. (1986) 'A standard predictive index of human response to the thermal environment', *ASHRAE Transactions*, vol 92, no 2, pp709–731

Humphreys, M. A. (1975) *Field Studies of Thermal Comfort Compared and Applied*, Current Paper CP 76/75, Department of the Environment, Building Research Establishment, Watford, UK (also in *Physiological Requirements on the Microclimate*, Institute of Hygiene and Epidemiology, Prague, 1975, pp115–181, and in *Journal of the Institution of Heating and Ventilating Engineers,* vol 44, 1976, pp5–27)

Humphreys, M. A. (1977) 'The optimum diameter for a globe thermometer for indoor use', *Annals of Occupational Hygiene*, vol 20, no 2, pp135–140

Humphreys, M. A. (1978) 'Outdoor temperatures and comfort indoors', *Building Research and Practice (J. CIB)*, vol 6, no 2, pp92–105

Humphreys, M. A. (1979) 'The influence of season and ambient temperature on human clothing behaviour', in Fanger, P. O. and Valbjorn, O. (eds) *Indoor Climate*, Danish Building Research, Copenhagen, pp699–713

Humphreys, M. A. (1981) 'The dependence of comfortable temperature upon indoor and outdoor climate', in Cena, K. and Clark, J. A. (eds) *Bioengineering, Thermal Physiology and Comfort*, Elsevier, London, pp229–250

Humphreys, M. A. (1995) 'Thermal comfort temperatures and the habits of Hobbits', in Nicol, F., Humphreys, M., Sykes, O. and Roaf, S. (eds) *Standards for Thermal Comfort*, Chapman & Hall, London, pp3–13

Humphreys, M. A. and Hancock, M. (2007) 'Do people like to feel "Neutral"? Exploring the variation of the desired sensation on the ASHRAE scale', *Energy and Buildings* (in press)

Humphreys, M. A. and Nicol, J. F. (1998) 'Understanding the adaptive approach to thermal comfort', *ASHRAE Transactions*, vol 104, no 1, pp991–1004

Humphreys, M. A. and Nicol, J. F. (2000) 'Effects of measurement and formulation error on thermal comfort indices in the ASHRAE database of field studies', *ASHRAE Transactions*, vol 106, no 2, pp493–502

Humphreys, M. A. and Nicol, J. F. (2002) 'The validity of ISO-PMV for predicting comfort votes in every-day life', *Energy and Buildings*, vol 34, pp667–684

Humphreys, M. A. and Nicol, J. F. (2004) 'Do people like to feel "Neutral"? Response to the ASHRAE scale of subjective warmth in relation to thermal preference, indoor and outdoor temperature', *ASHRAE Transactions*, vol 110, no 2, pp569–577

Humphreys, M. A. Nicol, J. F. and McCartney, K. J. (2002) 'An analysis of some subjective assessments of indoor air-quality in five European countries', *Indoor Air 2002*, vol 5, pp86–91 (*Proceedings of the 9th International Conference on Indoor Air Quality and Climate*, Monterey, California, July)

ISO (2005) ISO 7730: *Ergonomics of the Thermal Environment*, International Organization for Standardization, Geneva

Humphreys, M. A., Nicol, J. F. and Raja, I. A. (2005) 'Climate and comfortable indoor temperatures for the built environment – interpreting the worldwide data', in *1st International Conference, Environmentally Sustainable Development (ESDev-2005)*, 26–28 June, Abbottabad, Pakistan, vol 1, pp3–24

Kwok, A. G. (1998) 'Thermal comfort in tropical classrooms', *ASHRAE Technical Data Bulletin*, vol 14, no 1, pp85–101

Leaman, A. J. and Bordass, W. T. (1997) *Productivity in Buildings: The 'Killer' Variables*, Workplace Comfort Forum, London

Malama, A., Sharples, S., Pitts, A. C. and Jitkhajornwanich, K. (1998) 'An investigation of the thermal comfort adaptive model in a tropical upland climate', *ASHRAE Technical Data Bulletin*, vol 14, no 1, pp102–111

McCartney, K. J. and Nicol, J. F. (2001) 'Developing an adaptive control algorithm for Europe: Results of the SCATs project', in McCartney, K. (ed) *Moving Thermal Comfort Standards into the 21st Century*, Oxford Brookes University, Oxford, pp176–197

McIntyre, D. A. (1980) *Indoor Climate*, Applied Science Publishers, London

Nicol, J. F. (1974) 'An analysis of some observations of thermal comfort in Roorkee, India and Baghdad, Iraq', *Annals of Human Biology*, vol 1, no 4, pp411–426

Nicol, J. F. and Humphreys, M. A. (1973) 'Thermal comfort as part of a self-regulating system', *Building Research and Practice (J. CIB)*, vol 6, no 3, pp191–197

Nicol, J. F. and Humphreys, M. A. (2002) 'Adaptive thermal comfort and sustainable thermal standards for buildings', *Energy and Buildings*, vol 34, pp563–572

Nicol, J. F. and Humphreys, M. A. (2004) 'A stochastic approach to thermal comfort, occupant behaviour and energy use in buildings', *ASHRAE Transactions*, vol 110, no 2, pp554–568

Nicol, J. F. and Roaf, S. (2007) 'Adaptive thermal comfort and passive architecture', in Santamouris, M. (ed) *Advances in Passive Cooling*, James and James, London (in press)

Nicol, J. F. and Sykes, O. D. (1998) 'Smart controls for thermal comfort, the SCATs Project', *Proceedings of the EPIC 98 Conference*, Lyon, vol 3, pp844–849

Nicol, J. F., Raja, I. A., Alludin, A. and Jamy, G. N. (1999) 'Climatic variation in comfort temperatures: The Pakistan projects', *Energy and Buildings*, vol 30, no 3, pp261–279

Oseland, N. O. (1998) 'Acceptable temperature ranges in naturally ventilated and air conditioned offices', *ASHRAE Technical Data Bulletin*, vol 14, no 1, pp50–62

Pitts, A. (2006) 'The languages and semantics of thermal comfort', *Proceedings of NCEUB Conference*, Windsor

Poulton, E. C. (1973) 'Unwanted range effects from using within-subjects experimental designs', *Psychological Bulletin*, vol 80, pp113–121

Poulton, E. C. (1975) 'Observer bias', *Applied Ergonomics*, vol 6, no 1, pp3–8

Raja, I. A. and Nicol, J. F. (1997) 'A technique for postural recording and analysis for thermal comfort research', *Applied Ergonomics*, vol 27, no 3, pp221–225

Raja, I. A., Nicol, J. F. and McCartney, K. J. (1998) 'The significance of controls for achieving thermal comfort in naturally ventilated buildings', *Proceedings of the EPIC 98 Conference*, Lyon, vol 1, p63ff

Sharma, M. R. and Ali, S. (1986) 'Tropical summer index: A study of thermal comfort in Indian subjects', *Building and Environment*, vol 21, no 1, pp11–24

Shove, E. (2003) *Comfort, Cleanliness and Convenience: The Social Organisation of Normality*, Berg, Oxford

Sutter, Y., Dumortier, D. and Fontoynont, M. (2006) 'The use of shading systems in VDU tasks in offices, a pilot study', *Energy and Buildings*, vol 38, no 7, pp715–930

Wilson, M., Solomon, J., Wilkins, P., Jacobs, A. and Nicol, F. (2000) *Smart Controls and Thermal Comfort Project*, Final Report Task 1, University of North London, London

4

Typical Weather Years and the Effect of Urban Microclimate on the Energy Behaviour of Buildings and HVAC Systems

S. Oxizidis, A. V. Dudek and N. Aquilina

Abstract

Building simulation programs require climate information in the form of hourly based typical weather years assembled from meteorological measurements. A thorough description of these weather data sets is presented in this paper followed by a classification based on the way they are derived from monitored data and a comparison between the different types. Finally, we analyse the efforts made to deliver urban specific climatic data to assess the effect of urban microclimate on buildings.

■ *Keywords* – typical weather years; urban climate; building simulation

INTRODUCTION

Climatic data are of great significance in estimating a building's capability to protect its occupants from the outdoor climate and provide comfortable indoor climatic conditions. The amount of energy used in a building in order to smooth out the influence of the outdoor climate is a direct result of the microclimate around the building, among other factors. The lack of accurate climatic data could lead to bad design of buildings that either consume excess energy or cannot provide comfort.

Weather data files used in building simulation programs are generated from measured climatic data usually collected at remote meteorological stations, in most cases located out of urban centres (most often in nearby airports). No account is taken of the changes in meteorological conditions caused by the urban developments, yet it is known the differences are of major importance (Landsberg, 1981; Oke, 1988) and they significantly alter the energy behaviour of buildings and HVAC (heating, ventilation and air-conditioning) systems (Hassid et al, 2000; Papadopoulos, 2001; Santamouris et al, 2001).

The lack of urban specific climatic data leads to inappropriate design of urban buildings – not adapted to the urban microclimate – and wrong selection and sizing of

HVAC systems, burdening overall energy consumption in the building sector. In this paper, after a thorough description of the typical weather years (TWYs) used in energy calculations, a review of the efforts made to include the urban effect in weather data sets is presented.

DESCRIPTION OF WEATHER DATA FILES
CLIMATIC PARAMETERS AND TIME INTERVAL OF WEATHER DATA FILES

In principle, weather data for buildings simulation programs require climate information regarding the calculation of heating and cooling loads, in the form of time series of hourly values for a single year for several weather parameters, the most important of which are:

- air temperature;
- solar radiation;
- humidity; and
- wind speed.

In particular, these hourly data for a period of a year are supposed to represent climatic conditions considered to be typical over a long time period. Hence, these data, covering a whole year, are not an indicator of weather conditions to be met over the next year or the next five years, but they represent conditions estimated to be average over a long period of time like 15, 20 or 30 years.

The evolution of more sophisticated building simulation software nowadays allows the utilization of more detailed weather data both in time resolution and in the variety of climate parameters. This is because building simulation programs can now calculate a plethora of different aspects concerning the interaction of a building and the microclimate, including the determination of heating and cooling loads, energy consumption of HVAC systems, thermal and lighting comfort, and even indoor air quality. Thus, the need of additional climatic parameters emerges that previously did not exist or that were crudely calculated using other weather parameters.

The development of building technology itself leads simulation programs to become more demanding with respect to weather input parameters. The need for sub-hourly data emerges especially in cases where the construction of a building incorporates design aspects highly dependent on fluctuations of weather parameters. Such constructions may be lightweight building components that are very sensitive to solar radiation fluctuations, daylight design solutions depending greatly on changes in cloud conditions, or the use of photovoltaic cells where the relation with incident solar radiation is not linear, thus making inaccurate interpolated sub-hourly values derived from hourly observations (Hensen, 1999). This is of course true for most HVAC applications as well, where the system's capacitance is significantly lower than the building's capacitance.

An example of these new, more detailed, sets of weather data for building simulation programs is the new generalized weather data format that has been developed (Crawley et al, 1999) and adopted by two simulation programs, ESP-r and EnergyPlus. This weather file format can use sub-hourly data and a lot of additional climate parameters, though as yet some of the new parameters are inactive because they cannot be utilized by the corresponding simulation engines.

TIME PERIODS COVERED BY WEATHER DATA FILES

The weather data needed to determine the energy behaviour of a building and its systems should cover four distinct periods:

- a representative heating season;
- a representative cooling season;
- an extreme cold period; and
- an extreme warm period.

The first two periods are required in order to calculate the energy consumption of a building needed to cover the heating and cooling loads respectively, while the third and fourth periods are used for sizing the heating and cooling plant and equipment. Thus, there is a set of data known as design weather, representing severe climatic conditions, and another set known as typical weather, representing long-term climatic averages. Figure 4.1 shows the relationship between design and typical weathers with respect to energy calculations in buildings. The concept of design weather arises from the need for HVAC design calculations that estimate peak design loads based on selected indoor and outdoor design conditions.

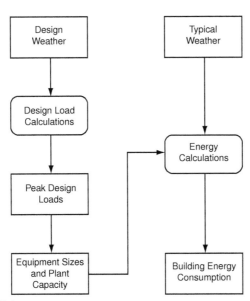

Source: Hui and Cheung (1997)

FIGURE 4.1   Design and typical weather

HISTORY AND TYPES OF WEATHER DATA FILES

Over the past three decades, several groups and organizations around the world involved in buildings and HVAC systems energy behaviour, and usually in cooperation with meteorological institutes, have developed weather data sets specifically designed for use in building energy simulations. Each data set was generated to serve a particular goal by following a specific method to determine which data from the actual weather data period of record would be used. The main differences, whenever these exist, and excluding the different statistical handling of recorded data, concern the components of solar radiation and the availability of measured data of solar radiation. Usually, the newer data sets are based on improved solar models or advanced solar radiation measurements.

It should be emphasized that the creation of these hourly data covering a whole year was exclusively carried out to serve energy calculation purposes initially for solar thermal systems and later for buildings as well. Thus, these typical weather years should not be considered as a climatological description of a specific location but as a tool for energy calculations. One can find a description and history of most of them, together with the current trends, in Crawley and Huang (1997), Hensen (1999), Chow et al (2006) and Chan et al (2006).

In principle, there can be a distinction between three different types of weather data for building simulation purposes, based on the way they derive from measured data. These are:

- historical years (HYs), for example, test reference year (TRY);
- typical years (TYs), for example, typical meteorological year (TMY); and
- synthetic years (SYs).

HISTORICAL YEARS

HYs are real data measured from a particular location. These are derived from a selection of a fixed continuous period (one year) of recorded data following some statistical rating to be used as reference. They are the earliest form of meteorological years for energy calculations in buildings.

A method for selecting the most representative year of a series of recorded years was used for the development of the TRY (NCDC, 1976). The TRY data refer to an actual historic year of weather, selected using a process by which years in the period of record that had months with extremely high or low mean temperatures were progressively eliminated until only one year remained. This tended to result in a particularly mild year since typical extreme conditions were excluded.

TRY contains dry bulb, wet bulb and dew point temperatures, wind direction and speed, barometric pressure, relative humidity, cloud cover and type, and a place holder for solar radiation, but no measured solar data. When used in building energy simulations, the simulation program typically estimates the amount of solar radiation based on the cloud cover and cloud type information available for the TRY location.

TYPICAL YEARS

TYs are artificial years assembled to match the long-term averages from a particular location using a particular statistical method. These data are created by selecting the most representative months (known as 'typical meteorological months' – TMMs) from long periods (for example, 30 years) of recorded data. So each resulting TY data file contains months from different years.

There are several statistical procedures to select the most representative months in order to shape a whole year. A description and a comparison of the most well-known with respect to their performance in energy calculations can be found in Argiriou et al (1999).

In establishing these typical weather years some statistical weighting is used to select which months are going to be included. This weighting is necessary in order to select the most representative months with respect to climatic variables important to the energy application to be used. So the months selected are average of the long-term climate only concerning these variables. This weighting usually favours temperature and solar radiation over humidity and wind.

For certain typologies of buildings that are 'defensive' with limited transparent surfaces and where the influence of temperature is dominant, the weather year should be developed using a higher weighting factor for the temperature and lower for solar radiation. For other 'aggressive' buildings with extended transparent surfaces, it may be solar radiation that is more deterministic; and for a third building types that incorporates natural ventilation techniques, the wind speed and wind direction may be crucial variables. Hence, a supposedly TY will somehow assume an average building whereas in reality there are many non-average buildings, each of which should ideally have its own TY (Hensen, 1999).

Extending this line of thinking, HVAC equipment need different weather years as well. An air-to-air heat pump with performance dominated by air temperature and to a lesser extent humidity will require energy calculations based on a weather year focused on these two variables, and a solar collector array will need data derived by statistical rating that favours solar radiation and to a lesser extent temperature and wind speed.

Table 4.1 presents the weighting factors used to select representative months during the creation of four different types of these weather years. The differences may not be large but still they are indicative of the effect of each climatic parameter on the energy performance of typical buildings.

The most ubiquitous weather year of this category is TMY (NCDC, 1981). TMY include all the climate parameters of TRY plus total horizontal and direct normal solar radiation. A similar format to the TMY was that used by ASHRAE to derive the WYEC (Weather Year for Energy Calculations) weather files (ASHRAE, 1985).

From the TY evolved TMY2 (Marion and Urban, 1995) and WYEC2 (ASHRAE, 1997). These newer data sets extended the previous format to include solar irradiation and illuminance data. In addition, they were based on newer measured data. Accordingly, ASHRAE developed the IWEC (International Weather for Energy Calculations) weather files (ASHRAE, 2002) that refer to international locations.

SYNTHETIC (STOCHASTIC) WEATHER YEARS

SYs are created by using several stochastic methodologies from long-term (monthly) averages and other statistics of the major weather variables (typically dry bulb temperature, humidity and global radiation) to generate time series of hourly based data.

TABLE 4.1 Weighting factors for the selection of representative months during the creation of typical weather years

PARAMETER	TMY	TMY2	CWEC (Canadian Weather for Energy Calculations)	CIBSE (Chartered Institute of Building Services Engineers) TRY
Max. dry bulb temperature	1/24	5%	5.0%	1/3
Min. dry bulb temperature	1/24	5%	5.0%	
Mean dry bulb temperature	2/24	10%	30.0%	
Max. dew point temperature	1/24	5%	2.5%	–
Min. dew point temperature	1/24	5%	2.5%	–
Mean dew point temperature	2/24	10%	5.0%	–
Max. wind velocity	2/24	5%	5.0%	1/3
Mean wind velocity	2/24	5%	5.0%	
Global solar radiation	12/24	25%	40.0%	1/3
Direct normal solar radiation	–	25%	–	

Source: Marion and Urban (1995); WatSun Simulation Laboratory (1995); Buhl (1999)

These SYs are used for locations where there is no availability of hourly recorded data. The procedures followed to generate these data are based on simplified models of the time series occurring in nature. These time series generators are just regressions, balance or thermodynamic equations (Aguiar et al, 1999). They generate each climatic variable by random choice using its statistical distribution (daily and hourly), autocorrelation and often taking into account its interactions with other variables (Adelard et al, 2000).

In general, synthetic (generated) weather data should respect the following principles (Adelard et al, 1997):

● the main statistic for each climatic variable (distributions laws, autocorrelation);
● the interactions between the climatic variables; and
● the influence of the geographic and physical environment on the climate of the specific site.

It should be noted that the accuracy of the models used in these weather generators is debated for regions and types of climate that are not well studied. In addition, wind speed and direction are roughly estimated due to their high dependency on local features.

SYs can be produced starting with input climatic monthly data only, widely available from meteorological atlases and bulletins. Usually, daily data are generated first and from these hourly data are derived. SYs can be made to yield exactly, at the output, these long-term monthly data.

Boland (1995) showed that synthetic series of climatic variables can be constructed with the same statistical properties as real time series for those variables at a specific location. In particular, he identified the significant cyclical components for solar irradiance and ambient temperature that enable the derivation of a methodology for the generation of synthetic series of daily values for both variables. Subsequently, he showed the procedure to generate hourly values from these determinations. These data series are statistically indistinguishable from the original data set and are claimed to be more representative than TYs since the methodology followed for their production guarantees that they represent long-term climatic averages.

There are several weather generators that create hourly weather data for a whole year using monthly averages as input. An extend list can be found in Skeiker (2005) for models that generate either individual climatic parameters like solar radiation and temperature, or the whole set of climatic variables needed in energy calculations.

The most popular and widely used are the models developed by Aguiar and Collares-Pereira (1988), Graham and Hollands (1990), Degelman (1991) and Knight et al (1991), on which most of the available weather generators are based. In addition, several other programs are available that are not so widely used or used particularly for specific climate types.

COMPARISON OF WEATHER DATA TYPES

So far it has been shown that users of building simulation programs have a vast variety of different weather files available to them – or the tools to create the files – from which they can choose the most suitable to predict the energy performance of a building and its systems. But the question arises of which is the best weather year.

The only criterion to estimate the worth of a TWY is how closely it represents the climatic long averages of a specific location with respect to energy performance simulations of buildings and their systems. Thus, the simulation output with the TWY as input should deviate as little as possible from the mean output obtained by using the complete multi-year data set from which the TWY was derived.

Argiriou et al (1999) compared several methodologies reported in the literature for the creation of TYS. They applied these methodologies to 20-year hourly recorded data from Athens, covering the period 1977 to 1996. Seventeen typical years were produced in total, which they used in simulations of solar systems. The results were cross-examined with the average output of the whole 20-year period. Due to the differences to the weighting and the statistical selection procedures the outcome was that, for every different application, another year performed better.

In a paper by Crawley and Huang (1997), different TYs were compared with a 30-year-long period of recorded data regarding their influence on simulated annual energy use, costs, annual peak electrical demand, heating and cooling load. For the simulations a test prototype office building was used, placed in five different US locations. The results showed that HYs are poorly representative of the long-term averages, while the newest sets of TYs that are based on improved solar models perform better.

Aguiar et al (1999) compared synthetic and typical years for Lisbon, Portugal, with respect to the thermal performance of a building test cell. They reported that all the synthetic data sets tested could substitute for the long-term observed series in a reasonable way, the details varying somewhat with the specific cell configuration and temperature control mode. In most cases the SY data were found to be the best alternative meteorological input. Particularly, when results were analysed at the monthly level, it appears that stochastic data were superior to classic TY data owing to their ability to exactly represent monthly mean meteorological values. For instance, for certain months, classic TY data can yield nearly perfect results when compared with long-term data, but for others they underestimate or overestimate significantly. With stochastic data the results are more regular, that is, underestimating or overestimating slightly at each month, and finally yielding somewhat better yearly performance estimates.

In addition, Boland and Dik (2001) argue that synthetically generated data sets are more useful for testing models of system performance than either short-term measured data or TMY data because, with an algorithm for generating synthetic data, one can produce any number of yearly data sets, the vast majority of which will exhibit characteristics representative of the long-term measured time series. Thus, one will have a multitude of possible input sets and this will enhance the testing because the performance can be judged on a variety of inputs, all typical of the location. By contrast, a TMY, even if representative of the location, will only provide one realization of the time series for testing the model.

Following another approach, Hui and Cheung (1997), based on current advanced computer's capabilities, suggest a multi-year simulation approach, the advantages of which are demonstrated with a 17-year simulation study for a commercial building in Hong Kong with a central HVAC system. In this way the disadvantages of using TMY are eliminated, though the higher computational effort is not insignificant.

By contrast, Degelman (1999) proposes the use of a typical week instead of the usual typical year simulation. Specifically, he used a carefully statistically generated week for each month, since a week is the minimum practical period replicated in a buildings' occupational and operational patterns and schedules. The results from simulations of a typical house in four US locations, having as reference output from TMY2 simulations, indicated much faster simulation time but a not small decrease in accuracy as well.

URBAN MICROCLIMATE AND ENERGY BEHAVIOUR OF BUILDINGS AND HVAC SYSTEMS

Recent rapid industrialization and urbanization have lead to an explosion in the number of urban buildings and consequently in energy consumption by this sector. An urban area, defined as a dynamic geographical morpheme with its own special features and processes, is characterized by an immense regional concentration of people, raw materials, production activities and energy consumption that cause a prominent perturbation on the local microclimate.

The influence of urban microclimatic conditions on the energy behaviour of buildings began to constitute a major research field in 1990s. The most perceivable impact of urban microclimates is higher air temperatures, a phenomenon known as the 'heat island' effect, the reasons for which are summarized by Oke et al (1991):

● The thermal and radiative properties (mainly increased heat capacity and decreased albedo) of the materials used in urban structures and covering urban land surfaces vary substantially from those of materials found naturally in rural areas.

- The lack of vegetation in the urban environment reduces transpiration as a source of latent heat loss. Extensive use of impervious surfaces and improved drainage also act to transport surface water quickly from the urban area, reducing latent heat loss through evaporation.
- The canyon-like topography of urban areas results in more effective absorption of short-wave radiation since it allows multiple reflections of solar radiation. Additionally, the multiple geometries allow for better absorption of sunlight during periods of high solar zenith angle, such as during sunrise and sunset.
- The canyon geometry decreases the efficiency with which the urban area can radiate long-wave radiation into the atmosphere and out into space. The multiple surfaces allow for the reabsorption of long-wave radiation, eliminating the loss of heat through radiative cooling.
- Urban areas also have an increased surface roughness, which slows down the surface winds. This decreases sensible heat loss from the urban surface through atmospheric convection.
- The atmosphere of urban areas typically has higher pollution levels than that of surrounding rural areas. Pollution, particularly aerosols, can produce a pseudo-greenhouse effect, absorbing and re-radiating long-wave radiation and inhibiting radiative surface cooling.
- Urban areas produce more heat by anthropogenic means than rural areas, due to the increased population density. Sources of anthropogenic heat include automobiles, construction equipment, air-conditioning units and heat losses from buildings.

Thus, the urban microclimate alters the values of several weather parameters, with significant effect on the energy performance of buildings, HVAC and renewable energy sources (RES) systems. This is why the estimation of an urban building's energy behaviour should at least take into account the modified values of the most important climatic parameters (temperature, solar radiation, wind speed and humidity).

In the literature two ways of meeting that challenge can be identified: first, the straightforward experimental approach that uses the results of field experiments that monitor climatic data inside the urban landscape, and second, the modelling approach that attempts to calculate the urban specific climatic data.

THE EXPERIMENTAL APPROACH

When following the conventional way to get TWY specific to a city (or even neighbourhood), an extensive network of *in-situ* weather stations is needed. Moreover, these stations should be in operation for an adequately long period of time to provide data that can result in the extraction of typical years to be representative of the long-term averages of the local microclimate. Hence, this is a very expensive procedure in both time and money. Thus, a very limited number of such studies have been conducted to date, and these focused only on air temperature (and sometimes wind speed) measurements.

Hassid et al (2000) estimated the effect of the Athens heat island on cooling energy consumption, both annual and peak, by using monitored data from 20 temperature stations in the Greater Athens region during the relatively cool 1997 and the relatively hot 1998 summers. The monitored data for these two years were compared to the TMY for Athens. Assuming the same humidity and solar radiation for all sites, they calculated the cooling demand of a four-apartment multi-family building during the cooling season months (May to September) using the building energy estimating software DOE2.1.E.

Santamouris et al (2001) present the results of monitored data collected for a three-year period from a network of measuring stations on the impact of increased ambient temperatures on the heating and cooling performance of buildings. Also, air flow and temperature distribution data derived by specific air flow and distribution experiments in ten urban canyons were used to evaluate the impact of canyon geometry and characteristics on the potential of natural ventilation techniques to provide passive cooling to urban buildings. Overall, three types of measurements were performed: air temperature, surface temperature and wind speed.

The collected data were used to calculate the distribution of the cooling and heating needs of a representative office building for all locations where climatic data were available. To study the performance of the building and to analyse in detail the obtained experimental data, a theoretical model of the building was created, the TRNSYS simulation tool.

In a paper by Kolokotroni et al (2006), climatic data (only for air temperature) over two weeks, one with extreme hot weather and one with typical hot weather in the centre of the London heat island and in a rural reference site, were used to carry out a parametric analysis on the energy demand for the cooling of a typical English office building by the use of the thermal simulation program, 3TC.

THE MODELLING APPROACH

Many projects have been carried out in the effort to determine the interaction between buildings and their surrounding microclimate using a modelling approach. However, most of the work has focused on the impact of the thermal behaviour of buildings on the urban climate and not on the impact of the microclimate on the building. In any case, because of the high computational effort needed by meteorological models, no climatic data in the time resolution used by building simulation programs could be derived. Instead, a simplistic thermal model of a building is usually developed to be combined with meteorological models and computational fluid dynamics (CFD) codes. With scarce exceptions, where the city specific weather is crudely estimated from background weather (the notn-urban sites where meteorological data are monitored), no discussion on TWY is included.

Williamson and Erell (2001) propose the use of a model to describe microclimatic conditions (air temperature) in the urban canopy layer based on previous work by Sharlin and Hoffman (1984) and Swaid and Hoffman (1990), which estimated the urban air temperature from prevailing regional background weather conditions expressed as a base temperature, the contribution of the solar radiation absorption on the surfaces of the urban canopy layer, and the cooling effect of the net outgoing long-wave radiation from the urban surfaces.

The newer model operates at hourly time intervals and beyond the above includes the dependence of total solar radiation absorption instead of direct radiation only, the modification of outgoing long-wave radiation loss by the inclusion of cloud cover, the thermal effect of vegetation in the urban site, and the contribution of anthropogenic heat. In addition, it incorporates better data handling and improved algorithms for shading, long-wave radiation and wind effects. The preliminary model results, compared with measured data, were promising as an alternative to adjust existing TWYs to include the urban effect.

A modelling approach with similar goals is also used by scientists to predict the impact of climate change on buildings. Researchers run atmosphere–ocean general circulation models similar to models used to predict the weather, in which the physics of atmospheric motion are translated into equations that can be solved on supercomputers that derive daily or monthly climatic values for different scenarios of climate change. The monthly or daily climatic values of future years are then used to create SYs for use in building simulation models so an assessment of the impact of climate change on buildings can be carried out.

Crawley (2003), running the atmosphere–ocean general circulation models for 16 combinations of scenario and climate prediction, determined changes in temperature, precipitation and cloud cover. The derived monthly average changes were used in a specific SY generation methodology to create new weather years for US location. With these SYs he managed to assess the impact of climate change on the energy performance of buildings using the EnergyPlus whole-building energy performance simulation software for a small office building in four US cities with different climatic characteristics.

In another study by Levermore et al (2004), synthetic years were generated based on daily climatic data derived by climatic models. Several statistical procedures for generating hourly data were manipulated in a way to utilize the specific variables of the climatic model while producing the necessary variables for the building simulation program. The effect of climate change on buildings was evaluated for a multi-storey office block in several UK regions.

In the same vein, Papadopoulos (2005) and Papadopoulos and Moussiopoulos (2004), within the ATREUS project, suggested a methodology – adopting a computational route – to assess the influence of urban microclimate on the energy behaviour of buildings and HVAC systems. This methodology is depicted in Figure 4.2. The idea on which the ATREUS methodology is based is to use a numerical weather prediction model (mesoscale model, in that case of the MM5 model) to obtain weather information at particular locations in a city. Ideally, for this approach, several years of meteorological data are needed to produce a proper climatology, which is, however, highly computer-time consuming. Instead of running the weather prediction model for several years, a surrogate can be provided by running a limited number of days (12 'typical days', one for each month), for a full year, and generating climatology from this data set.

The term 'typical day' is used to describe a day that has a climatic average close to – and thus representative of – the long-term averages of the months from which it was selected. If these typical days can be selected from a pool of a long period of records (using the same statistical procedures and weighting factors used in TY generation) they would be in place to provide to the model input data of mesoscale nature not influenced

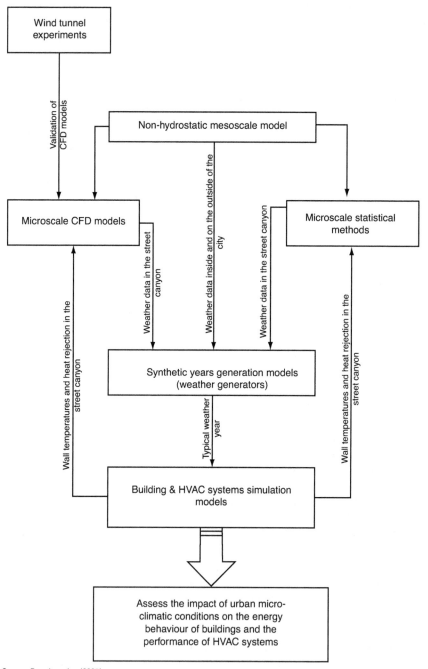

Source: Papadopoulos (2005)

FIGURE 4.2 The ATREUS methodology diagram

by the city. Thus, they can be representative at that scale. In parallel, by inserting in the model city specific characteristics such as land use, albedo of surfaces, ground moisture, anthropogenic heat produced and other factors that make the city climate different, a first order estimation of the city climate can be produced.

The resulting data from these simulations were used in three downscaling procedures: MM5 to building simulation program, MM5 to CFD models, and MM5 to statistical methods. MM5 data were used as boundary conditions for the CFD models, statistical methods and as direct input data to weather generators to produce a synthetic year for the building simulation models.

The CFD models were used to include the alterations in weather parameters imposed by the city geometry inside the urban canopy layer where mesoscale models cannot produce reliable data. Thus, by using the output of mesoscale modelling above the urban canopy layer as boundary conditions for microscale models, a more accurate calculation of the climatic data inside the city was possible. Indeed, this can be done to a very high resolution, even producing street specific climatic data.

The final output of the mesoscale models should deliver the exact variables and with the exact time resolution for building simulation models; however, a weather generator is also necessary to produce the proper climatic data. The only assumption needed for the use of these generators is that inter-variable relationships are the same for the urban climate as for the rural climate.

CONCLUSIONS

TWYs are a tool of great significance in the building simulation community. That is why they are continuously evolving, having in many cases advanced features like new climatic variables or sub-hourly intervals that still cannot be fully exploited by current simulation engines. But TWYs are still climatological descriptions of mainly rural locations near cities, although they are usually used for buildings located in urban areas.

None of the efforts made to assess the impact of specific urban climate on the energy performance of buildings has actually been derived by the creation of city specific TWYs. For this to be done networks of complete meteorological stations would need to be installed inside cities to monitor weather for at least a decade before real typical weather years could be produced.

Thus, a challenge emerges here for scientists to manage to adjust the existing TWYs for cities in order to include the urban effect by introducing the specific characteristics of each city.

AUTHOR CONTACT DETAILS

S. Oxizidis, Laboratory of Heat Transfer and Environmental Engineering, Department of Mechanical Engineering, Aristotle University of Thessaloniki, Greece

A. V. Dudek, Norwegian Institute for Air research, Kjeller, Norway

N. Aquilinia, School of Geography, Earth & Environmental Sciences, University of Birmingham, UK

ACKNOWLEDGEMENTS

The authors wish to thank the European Commission of Training and Mobility of Researchers Programme (DG Research) for funding the ATREUS project.

REFERENCES

Adelard, L., Boyer, H., Garde, F. and Gatina, J.-C. (2000) 'A detailed weather data generator for building simulations', *Energy and Buildings*, vol 31, pp75–88

Adelard, L., Garde, F., Pignolet-Tardan, F., Boyer, H. and Gatina, J.-C. (1997) 'Weather sequences for predicting HVAC system behaviour in residential units located in tropical climates', in *Proceedings of the 5th International IBPSA Conference Building Simulation*, Prague, www.ibpsa.org

Aguiar, R., Camelo, S. and Goncalves, H. (1999) 'Assessing the value of typical meteorological years built from observed and from synthetic data for building thermal simulation', in *Proceedings of the 6th International IBPSA Conference Building Simulation*, Kyoto, www.ibpsa.org

Aguiar, R. and Collares-Pereira, M. (1988) 'A simple procedure for generating sequences of daily radiation values using a library of Markov transition matrices', *Solar Energy*, vol 40, no 3, pp269–279

Argiriou, A., Lykoudis, S., Kontoyiannidis, S., Balaras, C.A., Asimakopoulos, D., Petrakis, M. and Kassomenos, P. (1999) 'Comparison of methodologies for TMY generation using 20 years data for Athens, Greece', *Solar Energy*, vol 66, no 1, pp33–45

ASHRAE (1985) *Weather Year for Energy Calculations*, American Society of Heating, Refrigerating, and Air-conditioning Engineers, Inc. Atlanta

ASHRAE (1997) *WYEC2, Weather Year for Energy Calculation 2, Toolkit and Data*, American Society of Heating, Refrigerating, and Air-conditioning Engineers, Inc. Atlanta

ASHRAE (2002) *International Weather for Energy Calculations (IWEC Weather Files) User's Manual, Version 1.1*, American Society of Heating, Refrigerating, and Air-conditioning Engineers, Inc. Atlanta

Boland, J. (1995) 'Time-series analysis of climatic variables', *Solar Energy*, vol 55, no 5, pp377–388

Boland, J. and Dik, M. (2001) 'The level of complexity needed for weather data in models of solar system performance', *Solar energy*, vol 71, no 3, pp187–198

Buhl, F. (1999) *DOE-2 Weather Processor*, LBNL Simulation Research Group, Berkeley, CA, http://gundog.lbl.gov/

Chan, A. L. S., Chow, T. T., Fong, S. K. F. and Lin, J. Z. (2006) 'Generation of a typical meteorological year for Hong Kong', *Energy Conversion and Management*, vol 47, pp87–96

Chow, T. T., Chan, A. L. S., Fong, S. K. F. and Lin, J. Z. (2006) 'Some perceptions on typical weather year – from the observations of Hong Kong and Macau', *Solar Energy*, vol 80, no 4, pp459–467

Crawley, D. B. (2003) 'Impact of climate change on buildings', in *Proceedings of the 2003 CIBSE/ASHRAE Conference*, www.cibse.org

Crawley, D. B., Hand, J. W. and Lawrie L. K. (1999) 'Improving the weather information available to simulation programs', in *Proceedings of the 6th International IBPSA Conference Building Simulation*, Kyoto, www.ibpsa.org

Crawley, D. B. and Joe Huang, Y. (1997) 'Does it matter which weather data you use in energy simulations?', *Building Energy Simulation News*, Simulation Research Group, Lawrence Berkeley National Laboratory, vol 18, no 1, pp2–12, http://gundog.lbl.gov/

Degelman, L. O. (1991) 'A statistically-based hourly weather data generator for driving energy simulation and equipment design software for buildings', in *Proceedings of the 2nd International IBPSA Conference Building Simulation*, Nice, Sophia-Antipolis, www.ibpsa.org

Degelman, L. O. (1999) 'Examination of the concept of using "typical-week" weather data for simulation of annualized energy use in buildings', in *Proceedings of the 6th International IBPSA Conference Building Simulation*, Kyoto, www.ibpsa.org

Graham, V. and Hollands, K. (1990) 'A method to generate synthetic hourly solar radiation globally', *Solar Energy*, vol 44, no 6, pp333–341

Hassid, S., Santamouris, M., Papanikolaou, N., Linardi, A., Klitsikas, N., Georgakis, C. and Assimakopoulos, D. N. (2000) 'The effect of the Athens heat island on air conditioning load', *Energy and Buildings*, vol 32, pp131–141

Hensen, J. L. (1999) 'Simulation of building energy and indoor environmental quality – some weather data issues', in *Proceedings of the International Workshop on Climate Data and their Applications in Engineering*, Prague, Czech Hydrometeorological Institute, www.bwk.tue.nl/

Hui, S. C. M. and Cheung, K. P. (1997) 'Multi-year (MY) building simulation: Is it useful and practical?', in *Proceedings of the 5th International IBPSA Conference Building Simulation*, Prague, www.ibpsa.org

Knight, K., Klein, S. and Duffie, J. (1991) 'A methodology for the synthesis of hourly weather data', *Solar Energy*, vol 46, no 2, pp109–120

Kolokotroni, M., Giannitsaris, I. and Watkins, R. (2006) 'The effect of the London urban heat island on building summer cooling demand and night ventilation strategies', *Solar Energy*, vol 80, pp383–392

Landsberg, H. E. (1981) *The Urban Climate*, Academic Press, New York

Levermore, G., Chow, D., Jones, P. and Lister, D. (2004) 'Accuracy of modeled extremes of temperature and climate change and its implications for the built environment in the UK', Tyndall Centre for Climate Change Research, Norwich, UK, www.tyndall.ac.uk/

Marion, W. and Urban, K. (1995) *User's Manual for TMY2s Typical Meteorological Years*, NREL/SP-463-7668, National Renewable Energy Laboratory, Golden, CO, http://rredc.nrel.gov/

NCDC (1976) *Test Reference Year (TRY), Tape Reference Manual*, National Climatic Data Centre, US Department of Commerce, Asheville, North Carolina

NCDC (1981) *Typical Meteorological Year User's Manual*, National Climatic Data Centre, US Department of Commerce, Asheville, North Carolina

Oke, T. R., Johnson, D. G., Steyn, D. G. and Watson I. D. (1991) 'Simulation of surface urban heat island under ideal conditions at night – Part 2: Diagnosis and causation', *Boundary Layer Meteorology*, vol 56, pp339–358

Oke, T. R. (1988) *Boundary Layer Climates*, Routledge, London

Papadopoulos, A. M. (2001) 'The influence of street canyons on the cooling loads of buildings and the performance of air conditioning systems', *Energy and Buildings*, vol 33, pp601–607

Papadopoulos, A. M. (ed) (2005) 'Advance tools for rational energy use towards sustainability with emphasis on microclimatic issues in urban applications', *Executive Scientific Report to the European Commission's Directorate-General for Research*, HUMAN POTENTIAL Programme, Research Training Networks, Contract No HPRN-CT-2002-00207

Papadopoulos, A. M. and Moussiopoulos, N. (2004) 'Towards a holistic approach for the urban environment and its impact on energy utilisation in buildings: The ATREUS project', *Journal of Environmental Monitoring*, vol 6, pp841–848

Santamouris, M., Papanikolaou, N., Livada, I., Koronakis, I., Georgakis, C., Argiriou, A. and Assimakopoulos, D. N. (2001) 'On the impact of urban climate on the energy consumption of buildings', *Solar Energy*, vol 70, pp201–216

Sharlin, N. and Hoffman, M. E. (1984) 'The urban complex as a factor in the air-temperature pattern in a Mediterranean coastal region', *Energy and Buildings*, vol 7, pp149–158

Skeiker, K. (2005) 'Mathematical representation of a few chosen weather parameters of the capital zone "Damascus" in Syria', *Renewable Energy*, vol 31, pp1431–1453

Swaid, H. and Hoffman, M. E. (1990) 'Prediction of urban air temperature variations using the analytical CTTC Model', *Energy and Buildings*, vol 14, no 4, pp313–324

WatSun Simulation Laboratory (1995) *Canadian Weather for Energy Calculations (CWEC files), User's Manual*, Environment Canada – Atmospheric Environment Service and the National Research Council of Canada, Waterloo, Ontario, www.cansia.ca/

Williamson, T. J. and Erell, E. (2001) 'Thermal performance simulation and the urban microclimate: Measurements and prediction', in *Proceedings of the 7th International IBPSA Conference Building Simulation 2001*, Rio de Janeiro, www.ibpsa.org

5

Energy Cost and its Impact on Regulating Building Energy Behaviour

Agis M. Papadopoulos

Abstract
The need to improve building energy behaviour was born out of the price shock caused by the oil crises in the 1970s. The response was expressed by national legislative acts regulating the demand for heating and ventilation. The results were important, though not always without side effects, for example, in the field of indoor air quality. Furthermore, economic and environmental considerations played an important role in determining the policies applied, the latter particularly in the 1990s and as a result of the Kyoto and Montreal Protocols. Finally, new problems, like the increasing demand for air conditioning and its impact on national electricity systems began to influence the way in which a building's energy behaviour is considered. The enforcement of the European Directive on the Energy Performance of Buildings (2002/91/EC) seems to provide for the first time an integrated regulatory tool, enabling the simultaneous consideration of the energy, environmental and economic parameters of buildings design. Its implementation, which is still facing delays, will prove the degree of its efficiency.

This paper discusses the evolution of these developments within the framework of energy regulation over the last 30 years, with specific examples from Europe. The discussion focuses on the thermal insulation of the building's envelope, the requirements for indoor air quality and the use of air conditioning, in order to narrow a subject that is too broad to be covered in its entirety. Developments from the first regulations in 1976 to Directive 2002/91 were neither straightforward nor solely driven by the rise of energy costs. They are based on the quest for an energy conscious, environmentally friendly and financially feasible building, which must also be friendly to its users.

■ *Keywords* – energy cost; policies; regulations; thermal loads; cooling demand

INTRODUCTION
The European Directive on the Performance of Buildings (2002/91/EC), the implementation of which has been mandatory for all member states since January 2006,

is the most recent in a long series of regulatory actions aiming at the improvement of building energy behaviour. This necessity to improve energy behaviour of buildings became peremptory during the two oil crises in the 1970s, and was expressed in the effort to reduce the demands for heating, ventilation and air conditioning (HVAC), without endangering the living standards of the day. However, the phenomenon is neither new nor one-dimensional. In the recent, but deceptively easily forgotten, first half of the 20th century, drastic energy conservation measures were applied affecting economic and social life in most European countries. As a result of two world wars, but also the Great Depression between them, energy became a precious commodity, though the building sector depended on coal rather than oil. At the end of the 20th century, and especially during the careless energy price era of the 1990s, most of the energy conservation actions taken, both on a national and on an international level, had their origins in environmental, rather than in purely energy saving, motives, an approach that is being reviewed due to the sharp increase in energy prices after 2003. Finally, the impact of establishing satisfactory indoor air quality conditions had, throughout the 20th century, a great impact on the ventilation of buildings, and hence on their energy behaviour. The main conclusion to be drawn from this brief and rather incomplete list of events and developments is that regulating the energy behaviour of buildings has been a goal that pre-dates the volatility of the energy markets (Figure 5.1) that have become the driving force of all actions since 1973.

In the 30 years of intensified, systemic development in the field of energy design of buildings, a new, interdisciplinary scientific field has developed, reaching a stage of maturity in a fairly brief time period. It is characterized by an advanced, and experimentally well

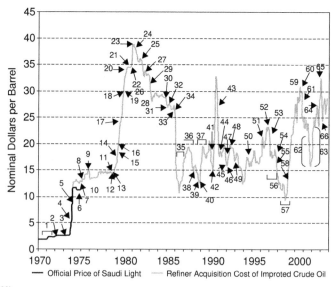

Source: EIA (2006)

FIGURE 5.1 Development of oil prices since 1970

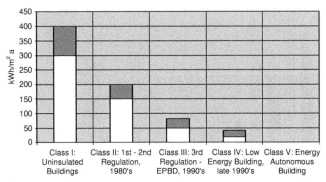

Source: Gertis (1999)

FIGURE 5.2 Evolution of heating energy consumption in German buildings, according to the five steps approach of Gertis

validated, theoretical background, by its incorporation in the syllabi of most engineering and architectural departments of universities and, at least in many countries, by a flexible and fairly effective legislative framework. At the same time, the architects, engineers and constructors active in the field have access to powerful computational codes, to new generations of insulating materials, building components, such as glazing, and HVAC systems, all of which enable the implementation of progressive solutions, besides having the effect of ensuring that less limitations are imposed on architectural design. These developments are exemplified in the evolution of building energy behaviour in countries like Denmark or Germany, where average specific annual consumption dropped from 300–400kWh/m²a in 1970 to less than 50kWh/m²a, according to Gertis's five steps classification (Figure 5.2) (Gertis, 1999). At the same time, in other European countries developments were not so spectacular. Taking as an indicator the average thickness of insulation enforced by national regulations in walls, two groups of countries can be identified: first, those that have progressively increased the required thickness by a factor of two (the Nordic countries and Germany being good example); and second, countries like Italy, Greece and Spain, where little has changed after the introduction of the first generation of regulations in the 1970s. This evolution, which is depicted in Figure 5.3, is discussed below, using the developments in Germany and Greece as examples. Still, whatever the motive of conservation actions may be, despite the progress made and probably because this progress has not been at the same pace throughout Europe, energy consumption in the building sector continues to constitute a major part of the worldwide annual final energy use. In the European Union alone it exceeds 40 per cent (European Commission, 2005).

FROM ENERGY CONSERVATION TO BUILDING PHYSICS: COPING WITH THE HEATING DEMAND

The first oil crisis in 1973 led to an increase in the price of crude oil by more than 300 per cent within less than two years. Given the fact that the importance of coal for thermal use in the building sector had been steadily declining since the 1950s and that the use of

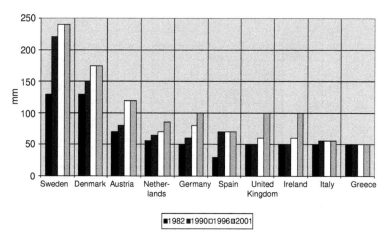

Source: EURIMA (2003); EUMEPS (2006)

FIGURE 5.3 Development of average insulation thickness for walls, as foreseen by national regulations, in European countries

electricity and natural gas was still rather negligible in Europe (though more popular in the US), the explosion in the retail price of heating oil affected entire societies (Hirst and Jackson, 1977). Reacting to this development, most European governments imposed regulatory policies on the building sector, aimed especially at new construction, mainly by means of setting tight limitations on the thermal transmissivity values of building envelopes, on reducing ventilation rates, and on increasing the minimum efficiency values to be achieved by new boilers. In that sense, the first German thermal insulation regulation of 1977, the WSVO 1977, is a good example of imposing limits on thermal losses through building envelopes and reducing ventilation/infiltration losses, by setting a maximum K-value with respect to a building's typological features, aiming at the reduction of values higher than $1.2W/m^2K$ (Papadopoulos, 1977).

A building's energy balance was considered largely in thermal terms. The policies applied were, at least quantitatively, successful and buildings constructed in the late 1970s and early 1980s showed significantly lower energy consumption values than their predecessors. In terms, however, of what is nowadays known as indoor environmental quality (IEQ) the results were not so encouraging. In the 1970s issues concerning indoor air quality (IAQ) were temporarily ignored, and the trend in construction was marked by increasingly airtight buildings and reduced ventilation rates. This resulted in ventilation requirements as low as 2.5L/s person (Janssen, 1999). In northern European countries, with harsh winter climate and very strict regulations, this led to impressive, albeit also dangerous, results as IAQ deteriorated in line with energy consumption reductions (Batty et al, 1984; Ihle, 1996). Similar trends were even monitored in countries with mild Mediterranean climate, as an analysis of typical Greek buildings demonstrated (Papadopoulos et al, 2002).

That the minimization of ventilation rates would lead to worsening IAQ was inevitable, but it was considered to be an acceptable risk, at least until the 1990s. Furthermore, the surface area of a building's openings was reduced in order to reduce transmissivity losses. Windows then rarely had K-values better than 3.5W/m²K, with the effect that natural lighting was reduced and the estrangement of the building's user to the environment could become a problem, especially in northern European countries (Schittich, 2003).

A second generation of regulations was produced in the 1980s. On the one hand, it featured tighter transmissivity values than previous regulations, but on the other hand, it enabled higher ventilation rates and, most importantly, it considered the building's envelope not only as a cause of thermal losses, but also a medium for possible thermal gains. The new regulations therefore signalled the transition from the enclosed, defensive building towards an open one, or what became known as the approach of building physics or bioclimatic architecture.

Capitalizing on the expertise gained during the 1970s, and from results obtained from large-scale applications such as in Milton Keynes in the UK, Solar Village III in Greece, or Horsens in Denmark, a systematic, more holistic approach was adopted towards the utilization of solar energy, natural lighting, natural ventilation, as well as the thermal storage properties of the building itself. Building Physics became an autonomous, interdisciplinary academic field; the first computational codes became available to architects and engineers, and new materials and techniques came on the market. Continuing to monitor the developments in Germany, the first thermal insulation regulation was updated in 1984 so that no building element was allowed to have a K-value exceeding 0.7W/m²K. In that way, and as the 1990s emerged and passed, the problem of coping with the heating loads of buildings was considered as a manageable one. A third update of the German thermal insulation regulation took place in 1995, limiting the K-values of the various building's elements to 0.5W/m²K. When combining the tighter legislation with the progress made in the field of thermal insulation materials, glazing and heating systems, then the positive results monitored in most of the European countries can be considered as a good example of the implementation of a sound energy conservation policy (Nadel et al, 2004; Papadopoulos, 2005).

The introduction in 1998 of the European Committee for Standardization (CEN) standard EN832, on the 'Thermal Performance of Buildings: Calculation of Energy Use for Heating Residential Buildings', marked a significant change of attitude; for the first time the term 'thermal performance' was used, instead of 'thermal protection' or 'thermal insulation', while the standard was not only limited to a static consideration of thermal losses, but also determined the final use of energy in order to cover the building's demands. The respective change in German regulations epitomized this approach, replacing the term 'thermal protection' with 'energy saving' (Reiss et al, 2002). Hence, one could argue, that the parties involved in the building sector were provided with an advanced, yet fairly comprehensible and flexible methodology to determine the thermal performance of buildings. Combined with the regulations applicable in most European countries, which enforced high standards (Figure 5.4), the building sector could be satisfied with the results, both in terms of energy savings achieved and with respect to the macroeconomics of the measures (Petersen and Togeby, 2001; Miguez et al, 2006). However, in the mid-1990s new challenges began to emerge.

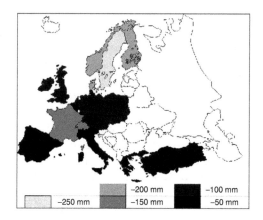

Source: EURIMA (2003)

FIGURE 5.4 Wall insulation thickness in European national regulations, 2003

FACING A 'NEW' PROBLEM: AIR CONDITIONING

In the beginning of the 21st century it became apparent that a new set of problems had come to dominate the field of building physics. In the 1990s cooling loads began to be a major component of a building's energy balance and, at the same time, a problem for electricity generation, transmission and distribution systems. An increased awareness of indoor air quality, due both to indoor and outdoor sources of pollution, led to the reconsideration of limitations imposed on ventilation (Vine, 2003). This interesting development in the minimum requirements imposed on ventilation is most clearly expressed in the changes of the relevant ASHRAE standards (Figure 5.5). Finally, with emphasis on the environmental quality of the building, not only in terms of living quality but with respect to its whole life cycle, the concept of life cycle analysis started to become attractive (Casals, 2006). Of these problems, coping with the cooling loads has become the most pressing, though it is closely linked to IAQ and building life cycles.

ESTABLISHING GOOD INDOOR ENVIRONMENTAL QUALITY IN A HOSTILE ENVIRONMENT

The influence of microclimatic conditions on the energy behaviour of buildings began to constitute a major research field in the 1980s. Increasing rates of urbanization, in both developed and developing countries, led to a drastic change in the geographical distribution of population and production activities, and consequently to a ubiquitous regional concentration of raw materials and energy consumption (WCED, 1987). The urban environment, characterized by a dense and often continuous layout of buildings, as well by as the use of materials with high thermal storage properties, contributes toward the appearance of the 'heat island' phenomenon causing air temperature increases in urban areas (Oke et al, 1991). The operational demands of buildings are both determined by this condition and contribute to its enhancement.

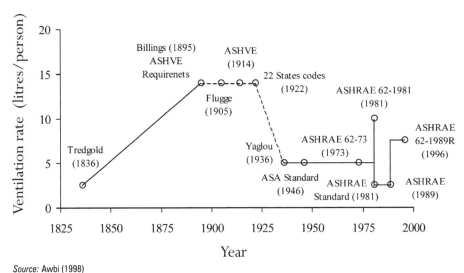

Source: Awbi (1998)

FIGURE 5.5 Development of minimum ventilation rates in the US

In that sense, the anthropogenic changes in urban microclimates and atmospheric pollution, combined with the increased awareness about the prevailing indoor air quality conditions due to pollutants emitted from building materials and anthropogenic activities, influence the demands on ventilation and air conditioning, leading to much higher energy loads than was the case up until the 1970s. It is clear that the whole concept and design of a building's HVAC system has to be considered on the basis of climatic and atmospheric conditions prevailing in contemporary urban areas, and not those measured at suburban airports, a practice established some decades ago (Santamouris et al, 2001).

Buildings ultimately have the goals of providing protection to human beings from the prevailing ambient conditions and supporting human activities, whether they are used either for residential or occupational purposes. Ventilation, be it natural or mechanical, has to achieve three main tasks: first, to remove air pollutants, generated by anthropogenic activities, from the interior by replacing the existing indoor air with fresh ambient air; second, to dilute the air pollutants in the interior by mixing the old and fresh indoor air; and third, to supply air to maintain good thermal comfort either by introducing colder air or by giving the necessary velocity to achieve the evaporative effect on the inhabitants. These three tasks are the basic design characteristics of any contemporary central HVAC system and can be achieved by any state-of-the-art system, however at a certain cost.

The acceptance of this cost is based on two facts. First, buildings constitute major capital investments and, especially in office buildings, several studies link IEQ not only to human health problems, but also to decreased productivity, highlighting indoor air climate as an essential quality of buildings (Wargocki et al, 1999). Second, human health is invaluable and the possibility of poor IAQ and/or poor thermal comfort conditions prevailing in building interiors contradicts their *raison d'être*. One has to keep in mind that the overwhelming majority of buildings in Europe feature neither mechanical ventilation nor central air-conditioning systems, leaving natural ventilation the only way to comply with IAQ requirements and, to a certain extent, to control indoor thermal comfort conditions. The implementation therefore of the proposed and published CEN standards, prEN 15242, prEN 13779, EN 13791 and EN 13792, which accompany the European Directive on the Energy Performance of Buildings and assess thermal comfort and IAQ in order to evaluate and classify mechanically and naturally ventilated buildings, constitutes a major step into the future (Railio, 2006).

ON THE MARKET AND ECONOMICS OF AIR CONDITIONING

For the reasons discussed above, since the late 1990s it has become commonly accepted that a lack of air conditioning in buildings designed without any particular bioclimatic features leads to unacceptably poor comfort conditions; an attitude that led to massive retrofitting of split-unit types of air-conditioning systems, also known as room air-conditioners (RACs). But even in buildings retrofitted in that way, the situation has not always improved in terms of IEQ. Because RACs do not enable the ventilation of interiors, the concentration of air pollutants may even increase in the building, underlining the point that increased energy consumption does not necessarily result in an improvement of indoor environmental conditions (Avgelis and Papadopoulos, 2003). The solution to this problem lies in natural ventilation carried out simultaneously with the operation of the RAC, leading obviously to even higher energy consumption values.

It is clear that the impact of increasing IEQ standards on energy consumption will become even more important in years to come, frequently exceeding the 'traditional' thermal losses that occurred in the heating period through the building's envelope. This has become clear in southern Europe, but is also beginning to reveal itself in central and northern Europe as the spread of air-conditioning appliances becomes the primary reason for increasing energy consumption in buildings, after consumption levels had declined due to falls in heating demands (Freedonia Group, 2002). RACs have become a major success story in the HVAC sector, winning market shares from central AC systems. While global air-conditioning sales are increasing at 4 per cent annually, sales of RACs are increasing at almost double that rate. In Greece the total installed electrical capacity of RACs is estimated at more than 3500MW, while approximately 200MW are installed every year. Similar developments can be monitored with respect to central or semi-central units: their capacity is estimated to be more than 1500MW with an annual growth of about 250MW, considering statistics for only the last five years.

Developments in southern Europe are in line with the Greek example: increased spread of RACs and their strong impact on the energy system. According to EERAC (1999), the penetration of RACs in southern European countries will increase by more than 10 per cent annually in forthcoming years, adding more loads to the electricity system and resulting in higher CO_2 emissions. Furthermore, the heat waves that hit western and northern Europe in 2003 acted as a catalyst for increased sales of RACs in markets that had hitherto been rather reluctant to adopt this technology. Combined with falling prices for mainly Asian produced RACs, sales are expected to increase by more than 20 per cent annually in western Europe, compared to 8 per cent for the year 2002 (RAC, 2003).

The use of RACs enhances the street canyon phenomena, a major factor in the development of heat islands in densely built urban areas. This is because the compressor units, which constitute emission points of rejected heat, are usually suspended on the buildings' façades or placed on the flat roofs. This rejected heat enhances, on a microscale level, the street canyon effect, increasing even more the cooling demand of the buildings. At the same time, it further reduces the coefficient of performance (COP) of the air-conditioners by up to 25 per cent due to the higher ambient air and surface temperatures, thus creating a vicious circle in terms of cooling and electricity demand (Hassid et al, 2000; Papadopoulos, 2001). These developments lead to an increase in electricity consumption that is concentrated in the two or three summer months, and become therefore a dominant element in the energy balance of every power supplier. As an example, the summer peak load in Greece during the years 1999–2000 showed an annual increase of 16 per cent or 1163MW, while the annual growth rate for load was approximately 4.1 per cent. At the same time, and for the years 1995–2000, the peak load demand increased from 5000MW to 8500MW (though it should be noted that until 1985 peak load demands were recorded in winter and not summer) (PPC, 2001).

The additional capacities needed to cover the peaks are obtained either by expensive generation facilities with low utilization factors like hydroelectric plants and gas turbine plants, or by electricity imports from neighbouring countries. Several times during recent years, the Greek electricity utility has faced the threat of blackouts on hot summer days, due both to limitations in production capacities and the overburdened distribution networks. Similar difficulties were experienced in Italy, Germany and France during the summer of 2003, despite the fact that on an annual base these countries show moderate growth rates in the use of electricity (Figure 5.6).

It is evident that such peaks, occurring only for a few weeks a year, cannot be covered at a reasonable cost because the additional investments needed in infrastructure cannot

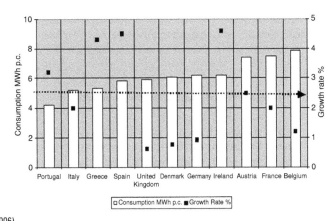

Source: EIU (2006)

FIGURE 5.6 Development of electricity consumption and its annual growth rate in European countries

be justified. In order to face the summer load peaks, without affecting the demand side, new power generation plants would have to be built close to consumption locations, that is, in the major urban areas, at a high economic and environmental cost. Alternatively, if the utility uses a time-dependent tariff or demand charge to cover the additional cost of the capacities needed, the operational costs of air conditioning would be much higher for customers. Such an alternative, which would discourage the extensive use of air conditioning, could only be efficient in the long run if it was accompanied by the introduction of energy conservation techniques to reduce cooling loads. But even ignoring the economics of such a policy, technical barriers emerge when attempting to cover a steeply increasing energy demand.

The example of California in 2000 demonstrates that there are clear limits to the increase of consumption that can be covered in a liberated energy market, even when demand side management tools are applied (Faruqui et al, 2001). The Californian example also indicates that when the total economic costs of covering the peak loads are taken into account, this may well lead to peak electricity retail prices three, five or even ten times higher than the respective base load prices; a cost that could not be carried even by the wealthy state of California. While lessons have been learned on how to manage the energy market in terms of an energy pool with typical commodity market's features, and at the same time on how to implement medium- and long-term capacity planning schemes, the necessity for reducing peaks in demand inevitably leads to adopting a reasonably differentiated pricing policy. According to studies made during the preparations for the Greek Olympic Games of 2004, the actual marginal cost for covering the peak summer time demand would well exceed €150/MWh, compared to approximately €42/MWh for base load production and €73/MWh for the summer period (RAE, 2004; CEPD, 2006) The results of these studies were, unfortunately, verified in the years 2004 and 2005 (RAE, 2005). Also, environmental or any other sort of externalities were not included in these figures, though they may very well have to be considered with respect to the obligations arising from the Kyoto Protocol and its impact on power generation. The evaluation of the energy design and behaviour of buildings, as well as of the conservation measures and/or various passive and active cooling and refrigerating systems, should be carried out with respect to this threshold. The main goals can only be to avoid or reduce the generation of cooling loads, to prevent the daily burdening of building interiors and, finally, to try to produce refrigeration in a sustainable way.

NON-ENERGETIC ASPECTS OF AIR CONDITIONING

The widespread use of air conditioning, however, is not only an energetic or economic problem. Neither are environmental problems limited to the CO_2 emissions due to energy consumption. Prior to the Montreal Protocol of 1987, the refrigeration and air-conditioning market was dominated by the use of chlorofluorocarbon (CFC) and hydrochlorofluorocarbon (HCFC) refrigerants. Their properties, like zero flammability and toxicity, make it relatively easy to design safe refrigeration systems that can be used in locations where untrained members of the public may be in the vicinity of a refrigeration plant. The European Commission's Regulation on ozone layer depleting substances (2037/2000/EC), which was implemented on 1 October 2000 is designed to control and phase out all ozone-depleting substances. Given the fact that a threat to the ozone layer from CFCs and HCFCs occurs only when these substances are released into the atmosphere, the EU regulation includes specific measures aimed at minimizing CFC and HCFC emissions, which have already led to an increase of the cost of owning and operating refrigeration systems using HCFCs, therefore supporting the emergence of either alternative refrigerants or alternative refrigeration technologies.

There are numerous new refrigerants on the market that have been specifically developed to address the phase out of CFCs and HCFCs. In the long run, however, only five important global refrigerant options remain for the vapour compression cycle, on which almost every contemporary air-conditioning system is based, as well as for the various non-vapour compression methods including sorption, steam jet, gas cycle cooling, and passive and natural cooling methods. These are: hydrofluorocarbons (HFCs) (HFC-blends with 400 and 500 number designation), ammonia (R-717), hydrocarbons (HCs) and blends (for example HC-290, HC-600, HC-600a), carbon dioxide (R-744) and water (R-718). None of these refrigerants is perfect. For instance, HFCs have relatively high global warming potential (GWP), ammonia is more toxic than the other options, and both ammonia and HCs are flammable. Interest in ammonia and hydrocarbons is stimulated, at least in part, by the fact that HFCs are greenhouse gases which will be, according to the Kyoto agreement, subjected to control measures. However, safety aspects also imply stringent emission controls for ammonia and HCs. Although these aspects are not covered by the Montreal Protocol, they nevertheless form criteria in the ongoing 'environmental acceptability' debate. Appropriate equipment design, maintenance and use can help to reduce these concerns, though at the cost of greater capital investment or lower energy efficiency. In turn, energy efficiency research is partly encouraged by the contribution of energy production to CO_2 emissions (UNEP, 2001). All these aspects transform future strategies on selection of refrigerant technologies into a delicate optimization procedure, at least on a medium- and long-term basis.

So far, the existing legislation on ozone depleting substances has placed increasing pressure on CFC and HCFC end-users to employ alternative working media and technologies. It has resulted in the extended use of HFCs, which are highly attractive for cooling applications. Despite the fact that all pure HFCs and most HFC blends require synthetic lubricating oils, instead of the more conventional mineral oils used with CFCs and HCFCs, the use of HFCs to replace CFC or HCFC in refrigeration plants is currently the option with the lowest cost for many users. By contract, while they have zero ozone depletion potential (ODP), HFCs have a significant GWP. This is typically in the range of 1000 to 3000 times the GWP of CO_2. At the 1997 Kyoto meeting, HFCs were included as one of six global warming gases being targeted for emission reductions. It is of interest to note that in 1995 the air-conditioning and refrigeration market was responsible for HCF emissions of 4.3 million tons of CO_2 equivalent, a figure corresponding to about 11 per cent of the total HCF emissions. This figure has been predicted to increase to 28.2 million tons of CO_2 equivalent by 2010 or approximately 43 per cent of total HCF emissions (March Consulting Group, 1998).

When considering alternative refrigeration technologies, solar refrigeration must be included, though this is not a new technology as such. Thermal driven absorption and adsorption technologies have been in use since the 1930s, while the use of solar driven sorption systems for the cooling of buildings has been studied systematically since the early 1970s. Despite some progress, however, no major commercial breakthrough has been achieved in the building sector, with the exception of desiccant systems, mainly in the US. Sorption systems refer to either open or closed cycles. Open cycles are mainly desiccant systems, while closed cycles are adsorption or absorption systems. Absorption systems are the oldest and most common heat driven systems and have been commercially available for many years. In the low-pressure side an evaporative refrigerant is absorbed to formulate a weak absorbent solution. Adsorption involves the use of solids for removing substances from either gaseous or liquid solutions (Papadopoulos et al, 2003). Both phenomena of absorption and adsorption are used to provide thermal compression of the refrigerant instead of mechanical compression in the case of vapour compression cooling systems. In desiccant systems, sorbents are used for the dehumidification of the incoming air, which is not strictly a refrigeration process but is certainly part of air conditioning.

By assessing the potential for solar energy refrigeration by means of a 'strengths, weaknesses, opportunities and threats' (SWOT) analysis, one can draw the following conclusions (Papadopoulos et al, 2004). Peaks in electricity demand occur, more frequently in recent years, in most developed countries during the summer period, for the reasons discussed above. The close coincidence of maximum insolation with both the cooling loads and the peak electricity demand indicates that solar-assisted refrigeration may be an interesting option to successfully handle these two issues. Furthermore, the solar thermal market has gained momentum in Europe since the mid-1990s, leading to a satisfactory spread of hot water systems, which may well also be used for solar cooling purposes. In terms of an emerging new market, this is the main argument in favour of partially solar powered cooling using alternative technologies like sorption or steam ejector cooling. Increasing the annual utilization factor of solar thermal systems, which is so far limited because of their use for hot water production and in some cases space heating, can counterbalance the lower efficiency of solar sorption systems and make them financially feasible options. At the same time they can be considered as the most environmentally friendly cooling option from all perspectives, including ODP, GWP and primary energy consumption. The most suitable combinations of solar thermal technologies and sorption systems are depicted in Figure 5.7.

TOWARDS THE REDUCTION OF COOLING LOADS AND DEMANDS

When determining a strategy, cooling loads in a building can be dealt with in three ways. First, by avoiding or reducing their generation in the first place through the application of the basic principles of building physics. This implies the implementation of sound sun-protection schemes, the use of thermal insulation, the use of reflective and low-absorbing materials on the building's interior, the practice of reasonable ventilation patterns, and the reduction of internal thermal loads production. Second, by postponing their impact on the building's interior. This presupposes a solid comprehension of the specific building's physics in order to capitalize on the building envelope's thermal storage capacity so as to delay the occurrence of high indoor air temperature values. In

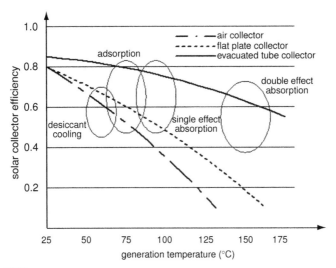

Source: Henning (2000)

FIGURE 5.7 Possible combinations of solar thermal and sorption refrigeration technologies

that way they can be dealt with late in the evening, when temperatures are cooler and cheaper load electricity tariffs apply. However, urban buildings, especially contemporary ones, are often light-weight constructions with low thermal storage capacity and it is difficult to apply this technique. Third, by using alternative sources and systems to produce the refrigeration necessary to cope with the cooling demand. This becomes necessary when the first and second options have been exhausted, or when they are impractical but when the problem of dehumidification demands that mechanical air-conditioning is essential.

Typically the 'passive' first and second options have been the paths favoured by architects and civil engineers, while the third 'active' option is favoured by mechanical engineers, traditionally with an industrial background. Still, the feasibility of the final option depends, to a large extent, on the successful implementation of the first two strategies and on the integrated energy design of the building. The idea of an exclusively naturally cooled building located in the densely built urban environment may seem attractive but it is hardly realistic. This is not only due to restrictions on the building architecture but also to the prevailing environmental conditions, that is, very low wind speeds and vortices in street canyons, high level of noise due to traffic, increased air pollution and so on.

In terms of thermodynamics, that is, thermal processes and systems design, the relationship between passive and active strategies can be expressed as one that ranges between covering the arising thermal load and ensuring the necessary installed power to cope with maximum demand. In terms of economics, this relationship can be described as the search for the minimum life cycle cost of the building, while at the same time trying to balance capital and operational expenses in a way that appeals to the building's owner or/and user. The approach adopted by the European Directive on the Energy Performance of Buildings, with the inclusion of renewable energy gains in the building's energy balance and the implementation of typified consumption values to be stated by the buildings' energy certificate, is an important step in that direction. It will enable the consideration of future operational expenses by the building's possible user, increasing the pressure on the constructor to abandon the typical criterion of minimizing the capital expenses by building an energy wasting building.

CONCLUDING REMARKS

The impact of increasing energy prices on the regulation of building energy behaviour has been the major driving force behind most of the legislative measures implemented by national governments and international organizations and institutions, at least in the aftermath of the turbulent 1970s. In the 1990s interest shifted towards facing environmental problems, both in terms of reducing atmospheric pollution and ensuring sustainability, and of establishing satisfactory indoor environmental conditions for building users. This was to some extent due to falling energy prices, but also to counterbalance some of the developments that arose in the 1970s, when the effort to reduce energy consumption caused serious drawbacks in the indoor environmental quality of buildings. However, the dilemma that appeared between energy consciousness and high standards of living quality is a false one, as developments since the mid-1990s have proved. Buildings can be simultaneously energy conscious, comfortable, healthy and environmentally friendly. At the same time, they can also be feasible investments, especially when considering the latest developments in the prices of oil, natural gas and electricity. In that sense, a building's life cycle cost and its environmental life cycle analysis indices can both be minimized, while at the same time ensuring a high quality of living or working environment.

In order to achieve this complex and ambitious goal the regulatory approach used hitherto with elaborated and strict, but fragmented, acts of legislation does not suffice. Three decades of intensive research in the field of building physics have proven that progress can only be achieved if the 'problem building' is considered through a holistic approach. The same philosophy has to be adopted in the regulatory approach, a philosophy that governs the new European Directive on the Energy Performance of Buildings, which provides a useful background for enforcing developments in the building sector. Key issues for its effective application are the use of state-of-the-art know-how, expertise and technology, the imposition of a firm set of standards and regulations in practice, and the elaboration of a new attitude among the market players. In that sense, the successful adoption by the market of the new Directive depends to a large extent on the adaptation of its philosophy in accordance with local needs, but also on the efficient training and re-education of those who will be asked to apply it. This need becomes more important in the light of a frequently vented criticism about the new Directive with its 31 accompanying standards. It is argued that it forms a complex problem for architects, engineers and constructors. This may be true to some degree, in particular when considering other regulations that already apply, such as those on refrigerants or others that are still being discussed, for example the minimum use of certain percentages of renewable energy sources directly for heating and cooling. However, complex problems can rarely be solved by simplified tools.

AUTHOR CONTACT DETAILS

Agis M. Papadopoulos, Associate Professor on Energy Systems, Laboratory of Heat Transfer and Environmental Engineering, Department of Mechanical Engineers, Aristotle University Thessaloniki, Thessaloniki, Greece
Tel: +30 2310 996011, fax: +30 2310 996012, email: agis@eng.auth.gr

REFERENCES

Avgelis, A. and Papadopoulos, A. M. (2003) 'The effect of ventilation on the indoor environmental quality in a natural ventilated office building, a case study', *Proceedings of 8th CEST*, Lemnos, Greece

Awbi, B. H. (1998) 'Ventilation', *Renewable and Sustainable Energy Reviews*, vol 2, pp157–188

Batty, W. J., O'Callaghan, P. W. and Probert, S. D. (1984) 'Energy and condensation problems in buildings', *Applied Energy*, vol 17, pp1–14

Casals, X. G. (2006) 'Analysis of building energy regulation and certification in Europe: Their role, limitations and differences', *Energy and Buildings*, vol 38, pp381–392

CEPD (2006) *Newsletters 21–25*, Center for Energy Policy and Development, University of Athens, Athens (in Greek)

EERAC (Energy Efficiency of Room Air-Conditioners) (1999) *Final Report of the EERAC Project*, SAVE Programme, EERAC, Armines, Paris, France

EIA (2006) *Annual Energy Outlook 2006*, Energy Information Administration, US Dept. of Energy, Washington DC, available at www.eia.doe.gov/oiaf/aeo/

EIU (2006) *Foresight 2020: Economic, Industry and Corporate Trends*, The Economist Intelligence Unit, available at http://newsroom.cisco.com/dlls/tln/research_studies/2020foresight

EUMEPS (European Association of Extruded and Expanded Polystyrene Manufacturers) (2006) *Reducing Climate Change with EPS Insulation*, EUMEPS, Brussels; available at http://www.eumeps.org/pdfs/reducing_climage_change.pdf

EURIMA (European Association of Mineral Wool Producers) (2003) *Thermal Insulation Thicknesses in Housing in Europe*, EURIMA, Brussels, www.eurima.org

European Commission (2005) *Green Paper on Energy Efficiency*, Directorate General for Energy and Transport European Commission, Luxembourg

Faruqui, A., Hung-Po, C., Niemeyer, V., Platt, J. and Stahlkopf, K. (2001) 'Analyzing California's power crisis', *Energy*, vol 22, pp29–52

Freedonia Group (2002) *World HVAC Equipment*, Cleveland, Freedonia Group Inc.

Gertis, K. (1999) 'How will we heat our houses in the future?', *Proceedings of the Forum Bau Leipzig, HTKW*, Leipzig, Germany (in German)

Hassid, S., Sanatamouris, M., Papanikolaou, N., Linardi, A., Klitsikas, N., Georgakis, C. and Assimakopoulos, D. N. (2000) 'The effect of the Athens heat island on air conditioning load', *Energy and Buildings*, vol 32, pp131–141

Henning, M. H. (2000) *Air-Conditioning with Solar Energy*, SERVITEC meeting, SERVITEC, Barcelona

Hirst, E. and Jackson, J. (1977) 'Historical patterns of residential and commercial energy uses', *Energy*, vol 2, pp131–140

Ihle, C. (1996) *Klimatechnik*, Werner-Verlag, Duesseldorf

Janssen, E. J. (1999) 'The history of ventilation and temperature control', *ASHRAE Journal*, September–October, pp47–52

March Consulting Group (1998) *Opportunities to Minimize Emissions of Hydrofluorocarbons (HFCs) from the European Union*, March Consulting Group, Manchester, UK

Miguez, J. L., Porteiro, J., Lopez-Gonzalez, L. M., Vicuna, J. E., Murillo, S., Moran, J. C. and Granada, E. (2006) 'Review of energy rating of dwellings in the European Union as a mechanism for sustainable energy', *Renewable and Sustainable Energy Reviews*, vol 10, pp24–45

Nadel, S., deLaski A., Kliesch, J., Shipley, A. M., Osann, E. and Harak, C. (2004) *Powerful Priorities: Updating Energy Efficiency Standards for Residential Furnaces, Commercial Air Conditioners, and Distribution Transformers*, ACEEE, Washington, DC

Oke, T. R., Johnson, G. T., Steyn, D. G. and Watson, I. D. (1991) 'Simulation of surface urban heat islands under "ideal" conditions at night: Part 2. Diagnosis of causation', *Boundary Layer Meteorology*, vol 56, pp339–358

Papadopoulos, A. M. (1977) *Thermal Insulation of Buildings,* Kyriakides, Thessaloniki (in Greek)

Papadopoulos, A. M. (2001) 'The influence of street canyons on the cooling loads of buildings and the performance of air conditioning systems', *Energy and Buildings*, vol 33, no 6, pp601–607

Papadopoulos, A. M. (2005) 'State of the art in thermal insulation materials and aims for future developments', *Energy and Buildings*, vol 37, no 1, pp77–86

Papadopoulos, A. M., Theodosiou, T. and Karatzas, K. (2002) 'Feasibility of energy saving renovation measures in urban buildings: The impact of energy prices and the acceptable pay back time criterion', *Energy and Buildings*, vol 34, pp455–466

Papadopoulos, A. M., Oxizidis, S. and Kyriakis, N. (2003) 'Perspectives of solar cooling in view of the developments in the air-conditioning sector', *Renewable and Sustainable Energy Reviews*, vol 7, no 5, pp419–438

Papadopoulos, A. M., Oxizidis, S., Doukas, H. and Samlidis, I. (2004) 'Constraints and potential for the propagation of solar refrigeration in the building sector', *Ktirio*, vol A-B, pp41–50 (in Greek)

Petersen, S. L. and Togeby, M. (2001) 'Demand for space heating in apartment blocks: Measuring effects of policy measures aiming at reducing energy consumption', *Energy Economics*, vol 23, pp387–403

PPC (1988–2001) *Annual Technical Bulletins*, for the years 1988–2001, Athens, Hellenic Public Power Corporation (in Greek)

RAC (2003) 'Four years after: The RAC story', *Eurovent/Cecomaf Review*, no 57, pp8–10

RAE (Regulatory Authority of Energy) (2004) *Synoptic Report on the Situation of the Electricity Market*, RAE, Athens (in Greek)

RAE (2005) 'Internal report on the electricity sector', RAE, Athens (in Greek)

Railio, J. (2006) 'EPBD: Influences on European standardization and on ventilation and air-conditioning industry', *REHVA Journal*, vol 1, pp19–23

Reiss, J., Erhorn, H. and Reiber, M. (2002) 'Energy renovated residential buildings: Measures – energy savings – costs', IRB Verlag , Stuttgart (in German)

Santamouris, M., Papanikolaou, N., Livada, I., Koronakis, I., Georgakis, C., Argiriou, A. and Assimakopoulos, D. N. (2001) 'On the impact of urban climate on the energy consumption of buildings', *Solar Energy*, vol 70, pp201–216

Schittich, C. (2003) 'From passive utilisation to a smart solar architecture', *Solares Bauen*, Birkhäuser, Edition Detail, pp13–25 (in German)

UNEP (United Nations Environment Programme) (2001) *Protecting the Ozone Layer – Volume 1: Refrigerants*, UNEP, Nairobi

Vine, E. (2003) 'Opportunities for promoting energy efficiency in buildings as an air quality compliance approach', *Energy*, vol 28, pp319–341

Wargocki, P., Wyon, P. D., Baik, K. Y., Clausen, G. and Fanger, P. O. (1999) 'Perceived air quality, sick building syndrome (SBS) symptoms and productivity in an office with two different pollution loads', *Indoor Air*, vol 9, pp165–179

WCED (1987) *Our Common Future*, World Commission on Environment and Development, Oxford University Press, Oxford

FURTHER LITERATURE

ASHRAE (American Society of Heating, Refrigerating and Air-conditioning Engineers) (1981) *ASHRAE Standard 55-1981: Thermal Environmental Conditions for Human Occupancy*, ASHRAE, Atlanta

ASHRAE (2001) *ASHRAE Standard 62-2001: Ventilation for Acceptable Indoor Air Quality,* ASHRAE, Atlanta

Brager, G. and de Dear, R. (1998) 'Thermal adaptation in the built environment: Literature review', *Energy and Buildings*, vol 17, pp83–96

ECA (European Collaborative Action on Urban Air, Indoor Environment and Human Exposure) (2003) *Ventilation, Good Indoor Air Quality and Rational Use of Energy*, report no 23, EUR 20741 En, Office of Official Publications of the European Communities, Luxembourg

Elagöz, A. (1994) 'Legal and administrative measures for energy savings in non-industrial buildings in the member states of the EC', *Renewable Energy*, vol 4, no 1, pp109–112

European Commission (2000) *Regulation (EC) No 2037/2000 of the European Parliament and of the Council of 29 June 2000 on Substances that Deplete the Ozone Layer*, OJEC, L244,29/09/2000, *European Commission*, Luxembourg

European Commission (2003) *Directive 2002/91/EC of the European Parliament and of the Council of 16 December 2002 on the Energy Performance of Buildings*, OJEC, L1/65,04/01/2003, European Commission, Luxembourg

European Committee for Standardization (1998) *EN 832 on the Thermal Performance of Buildings: Calculation of Energy Use for Heating Residential Buildings*, European Committee for Standardization, Luxembourg

Hegner, H. D., Hauser, G. and Vogler, I. (2005) *EnEN Novelle, Bauphysik spezial*, Verlag Ernst & Sohn, Berlin (in German)

Höppe, P. and Martinac, I. (1998) 'Indoor climate and air quality: Review of current and future topics in the field of ISB study group 10', *International Journal of Biometeorology*, vol 42, pp1–7

Norris, G. (2001) 'Integrating life cycle cost analysis and LCA', *International Journal of LCA*, vol. 6, no 2, pp118–121

Rehm, J. (2005) *Energy Saving Regulation with Accompanying Norms*, Verlagsgruppe Huethig-Jehle-Rehm, Berlin (in German)

Santamouris. M. (ed) (2000) *Energy and Climate in the Urban Built Environment*, James and James, London

6

Heat Island Research in Europe: The State of the Art

Mat Santamouris

Abstract

This paper summarizes recent research on urban heat islands carried out in Europe. Recent studies to identify the amplitude of urban heat islands as well as the main reasons that increase temperatures in urban areas are identified and discussed for southern, mid- and northern Europe. Studies aiming to identify the energy impact of heat islands are presented, as well as those that explore the impact of heat islands on the cooling potential of natural and night ventilation techniques. New deterministic and data-driven models developed to estimate the amplitude of heat islands are discussed and the results are presented of recent research on mitigation techniques and, in particular, the impact of green spaces, as well as the development and testing of white and coloured cool materials.

■ *Keywords* – heat island; energy in buildings; urban environment

INTRODUCTION

Heat islands are one of the best documented phenomenon of climate change. Heat islands have been the subject of research in Europe for at least 100 years. It is beyond the scope of this paper to to review all this research and present an historical analysis, and instead the paper concentrates on the progress achieved over the last 15 years. Studies on ambient air heat islands are presented for southern, mid- and northern Europe. Research based on remote sensing techniques are not considered. The main climatic conditions associated with the development of urban heat islands, as generalized by Oke (1982) and recently summarized by Arnfield (2003), are presented for some European cities.

Heat islands have a very important impact on the energy consumption of buildings (Akbari et al, 1992). Increased urban temperatures exacerbate the cooling load of buildings, increase the peak electricity demand for cooling and decrease the efficiency of air-conditioners (Santamouris, 2001; Santamouris et al, 2004). In parallel, high urban temperatures considerably decrease the cooling potential of natural and night ventilation techniques and increase pollution levels. Important studies on the energy impact of urban heat islands (UHIs) as well as their impact on passive cooling techniques and pollution levels have been carried out recently and are summarized below.

Modelling of the heat island phenomenon is a subject that has attracted many researchers. New deterministic and data-driven techniques have been developed, including models based on the use of neural networks that predict the amplitude of a heat island as a function of the main climatic parameters. The main characteristics of the newly developed models are also presented below.

Various mitigation techniques to decrease the impact of heat islands have been proposed. Among the most important techniques are the use of green spaces and cool sinks, as well as the extensive use of cool materials. Appropriate materials used in the urban scale can decrease surface and ambient temperatures, improve thermal comfort and decrease cooling loads of buildings. Extensive research described by Akbari et al (2005) has enabled the development of cool materials that present high solar reflectivity combined with high emissivity coefficients. White and coloured cool materials have been recently developed and tested in southern Europe, presenting very interesting optical and thermal performances. The paper reviews the progress achieved and the main results.

HEAT ISLAND INTENSITY OF EUROPEAN CITIES

Important heat island studies have been performed in Europe over the last 15 years. The work is based either on statistics of temperature differences between pairs or groups of urban and rural stations (Eliasson and Holmer, 1990; Yaggie et al, 1991; Bacci and Maugeri, 1992; Moreno-Garcia, 1994; Eliasson, 1996; Shahoedanova et al, 1997; Tayanc and Toros, 1997; Bohm, 1998; Klysik and Fortuniak, 1999; Beranova and Huth, 2005), on results obtained by networks of fixed stations in a city (Karaca et al, 1995; Dupont et al, 1999; Lazar and Podesser, 1999; Tumanov et al, 1999; Santamouris, 2001; Watkins et al, 2002a; Cristen and Vogtt (2004); Szymanowski, 2005), or by using mobile stations across an urban area Eliasson, 1996; Unger, 1996; Pinho and Manso Orgaz, 2000; Montavez et al, 2000; Alcoforado, 2002; Bottyan and Unger, 2003; Szymanowski, 2005; Szegedi and Kircsi, 2003).

Most of the studies are organized around the main generalization offered by Oke (1982) and summarized by Arnfield (2003). The impacts of wind, cloud cover and generally of cyclonic or anticyclonic conditions on the intensity of heat islands are reported. Also the time period of heat islands is presented and many studies discuss the season when UHIs reach their maximum intensity. Studies have been performed either in the Mediterranean area, central or northern Europe. The main experiments and the corresponding results are reported below and in Table 6.1.

MEDITERRANEAN ZONE

Alcoforado (2002) performed heat island measurements in Lisbon, Portugal, using mobile stations in at least 20 sites of the city between 4 January and 15 January 1995. It was found that the nocturnal air UHI occurs both in winter and summer nights with a mean intensity close to 2.5°C. In another study by Alcoforado and Andrade (2006), based on 69 air temperature measurement sites, the authors found that maximum UHI intensity is close to 3.5°C and the islands develop in the most densely built-up city districts.

TABLE 6.1 Results of heat island studies of European cities

CITY	IMPACT OF WIND	IMPACT OF CLOUD COVER	IMPACT OF CYCLONIC AND ANTICYCLONIC CONDITIONS	PERIOD THAT THE UHI DEVELOPS
SOUTHERN EUROPE				
Athens, Greece	Heat island intensity is increasing under a high pressure ridge that is characterized by weak pressure gradient and weak, variable winds or calms		High pressure ridge and closed anticyclone conditions characterized by the presence of a closed anticyclone accompanied by weak winds from the southern and northern sector	Summer period
Rome, Italy	The wind action, even if not covering the effect of the UHI, determines a reduction of UHI differences with respect to those that could be recorded in the case of calm wind conditions			The heat island clearly appears either during the winter months or during the summer months with growing values from the winter toward the summer, except for December when a second maximum is found
Parma, Italy				The difference varies seasonally with the maxima in spring and summer
Florence, Italy	Maximum heat island intensity on calm days	Maximum heat island intensity on clear days		Summer period
Lisbon, Portugal	Weather types with northerly winds are associated with the highest air temperature in the downtown area, partly because of a shelter effect	Analysis performed only for cloudless nights		The seasonal variation of UHI frequency showed a maximum in wintertime

TABLE 6.1 Results of heat island studies of European cities (Cont'd)

CITY	IMPACT OF WIND	IMPACT OF CLOUD COVER	IMPACT OF CYCLONIC AND ANTICYCLONIC CONDITIONS	PERIOD THAT THE UHI DEVELOPS
Aveiro, Portugal	Intensity of the island at its maximum when there is no wind	Intensity of the island at its maximum when the sky is totally clear	Weak UHI are associated with low pressure (cyclonic) or perturbation – atmospheric instability, strong winds and cloudiness, the occurrence of precipitation. The high-intensity islands correspond to high-pressure (anticyclonic) situations – clear sky and no wind	
Madrid, Spain	Higher UHI correspond to low wind speeds	Intensity of the heat island at its maximum when the sky is totally clear	Classification of UHI values on the basis of different types of weather shows that the highest values correspond to anticyclonic situations during the cold period, and that the lowest correspond to other situations occurring during this period	Maximum at summer time
Granada, Spain	For low winds situations, the maximum differences are higher	For clear sky situations, the maximum differences are higher		Maximum differences occur during winter months
Izmir, Adana, Bursa and Gaziantep, Turkey				Urban warming is more or less equally distributed over the year with a slight increase in the autumn months

CENTRAL EUROPE AND UK

Bucharest, Romania	The highest UHI is under low wind speeds	The highest UHI is under clear skies		Heat islands develop most in spring and summer, rather than in winter anticyclone situations
Szeged, Hungary	Calm or slight wind favourable for a strong development of the heat island effect	Little or no cloud coverage favourable for strong development of the heat island effect	Anticyclonic weather situations favourable for strong development of the heat island effect	
Debrecen, Hungary	Strong heat islands developed under anticyclonic weather conditions with weak wind speeds	Strong heat islands developed under anticyclonic weather conditions with clear skies and weak winds	Under anticyclonic conditions strong heat islands formed, but their shape was usually deformed by the prevailing winds. Strong cyclonic activity eliminated the formation of the UHI, while under weak cyclonic activity, regular but weak heat islands developed	In the non-heating season stronger heat islands develop than in the heating season
Wroclaw, Poland	An increase of wind speed to over 4m/s at night, and over 1m/s during the daytime, irrespective of cloudiness, totally eliminates the UHI or causes a considerable reduction in its intensity ($<1°C$)	The impact of cloudiness is practically unnoticeable during the daytime. At night only an increase in cloudiness to greater than 6 oktas is seen to diminish the UHI intensity		The highest average UHI intensity values are observed in the warm season, mainly in spring (May, April)

TABLE 6.1 Results of heat island studies of European cities (Cont'd)

CITY	IMPACT OF WIND	IMPACT OF CLOUD COVER	IMPACT OF CYCLONIC AND ANTICYCLONIC CONDITIONS	PERIOD THAT THE UHI DEVELOPS
Lodz, Poland	At night the UHI reaches its largest intensity in windless and cloudless conditions	The greatest differences occur during summer nights when skies are clear	Exceptionally intense UHIs that last for most of the night are associated with the advection of cold arctic air in which an anticyclone develops	
Prague, Czech Republic			Increase in heat island intensity is steeper under anticyclonic than cyclonic conditions in all seasons except for spring	UHI increases in all seasons as well as annually; the increase is significant in all seasons except for winter when the trend is much weaker
London, UK				Measurements only in summer time
NORTHERN EUROPE				
Moscow, Russia	Synoptic analysis of the period between May and August confirms that low wind speed conditions associated with anticyclones generate strong heat islands	Clear sky conditions associated with anticyclones generate strong heat islands		The highest intensities of heat islands were observed between May and August
Gothenburg, Sweden	Heat island intensity is at least 2.5°C, when the wind speed is less than 3m/s	The heat island intensity is at least 2.5°C, when the sky is clear	Measurements performed only under anticyclonic conditions	

Pinho and Manso Orgaz (2000) measured heat island intensity in the coastal city of Aveiro, Portugal, using mobile surface stations. Measurements were taken during 48 nights in the summer, autumn and winter of 1996, between 23.00 hours and 01.00 hours. They found that heat island intensity can reach 7.5°C and that the anticyclonic conditions of clear sky and calm weather accentuate heat islands while cyclonic conditions lessen their intensity.

Balkestahl et al (2006) took measurements of ambient temperature in Oporto, Portugal, from November 2003 to January 2005, using a mobile station. It was found that maximum heat island intensity is close to 7.3°C and occurs mainly during periods of low wind speed.

Yaggie et al (1991) analysed temperature time series from three rural and one urban stations in Madrid, Spain, for the period between 1965 and 1987. UHI intensity shows a minimum value in spring and a maximum in summer (3.1°C), and regarding daily minimum temperatures it was also found that the highest values correspond to anticyclonic situations during the cold period and that the lowest correspond to cyclonic conditions.

The UHI of Rome was initially studied by Colacino and Lavagnini (1982) using a network of ten urban and rural stations. By comparing measurements of different urban stations against a rural station for a 12-year period (1964–1975), they found that heat island intensity, based on the minimum temperatures, was close to 2.5°C during the winter and 4.3°C during the summer. Results of a simulation study reported by Bonacquisti et al (2006) show that UHIs are a nocturnal phenomenon, present both in winter, when the greatest difference between urban and rural temperatures is about 2°C, and in summer, when the temperature difference is about 5°C, mainly resulting from the urban geometry and the thermal properties of materials.

Zanella (1976) performed heat island measurements in the medium-sized city of Parma, Italy (170,000 inhabitants). Comparison of urban and a rural (airport) data for the period 1959–1973 showed an average difference of 1.4°C, with the maximum difference of 1.6°C occurring when daily temperatures reached their maximum. Zanella also reported that the difference varied seasonally, especially in spring and summer. Also in Italy, Bacci and Maugeri (1992) took measurements in Milan (1,600,000 inhabitants) and found that the temperature difference between the city centre and Linate airport was close to 1.4°C, with a heating rate of 0.13°C per decade.

In Florence, Petralli et al (2006) used surface temperature stations to measure ambient temperatures during the summer of 2005. The heat island intensity was found to be close to 3°C, with maximum intensity on clear and calm days.

In Spain, Moreno-Garcia (1994) collected data in Barcelona and found that heat island intensity reaches a maximum of nearly 8°C. While in Granada, Montavez et al (2000) examined the UHI intensity using stationary and mobile stations. The results show that the UHI phenomenon is stronger in winter and the maximum difference occurs in early morning when temperatures are at their daily minimum. The measured intensity of the heat island was found to be nearly 5°C.

Tayanc and Toros (1997) studied heat islands in the cities of Izmir, Adana, Bursa and Gaziantep in Turkey. Data from an urban and a rural temperature station in each city were used for comparison and cover the period 1951–1990. The analysis shows that there is a shift towards the warmer side in the frequency distributions of daily minimum and 21.00 hour temperature difference series. Seasonal analysis of individual 21.00 hour temperature series suggests that the regional warming is strongest in spring and weakest in autumn and winter. The intensity of heat island is found to range between 6.5°C and 9°C in the four cities. Urban warming was detected to be more or less equally distributed over the year with a slight increase in the autumn months.

Karaca et al (1995) studied the heat island effect and the impact of urbanization in Istanbul and Ankara in Turkey. Multi-year minimum temperature differences from seven urban and rural stations for Istanbul show a tendency for a climatic change towards a stronger UHI intensity. It is concluded that the urbanization effect is predominantly a night-time phenomenon for Istanbul where the maximum UHI effect is 0.0297°C and the temperature differences between urban and rural stations are close to 2°C. For Ankara, analysis of urban temperatures does not show any warming trend.

Santamouris (2001) reports measurements of heat islands in the Greater Athens area, Greece. Almost 30 surface temperature stations were used and measurements taken on an hourly basis for many years. It is reported that maximum heat island intensity in the central area is around 16°C, with a mean value for the larger central zone being 12°C. Also, absolute maximum temperatures in the central area are close to 15°C higher than in the suburban areas, while absolute minimum temperatures are up to 3°C higher in the centre. It was also found that heat island intensity in the central Athens park area is 6.1°C, while the heat island intensity of nearby located stations is 10°C.

In a further study reporting the results of the heat island research in Athens, Livada et al (2002) found that, in places near the sea, air temperatures are higher in the cold period of the year, not because of urbanization but mainly due to the influence of the sea, which favours the maintainance of high air temperatures. It is also reported that the persistence of high air temperatures during the hot period of the year or low air temperatures in the cold period is mostly related to synoptic weather conditions and cannot reasonably be considered as an index for heat island development.

In another study by Mihalakakou et al (2002), aiming to investigate the impact of synoptic scale atmospheric circulation on the UHI over Athens, it is reported that synoptic scale circulation is a predominant input parameter with a considerable effect on the heat island intensity. It was also found that high pressure ridges encourage the heat island phenomenon, while under weather conditions characterized by intense northern winds, heat islands do not appear.

Dandou et al (2005) also studied the heat island effect and the impact of urbanization in the Greater Athens area. They found that the urbanized version of the MM5 mesoscale model calculated a strengthening of the nocturnal urban heat island. Changes in the air temperature at four stations proved to be favourable through the whole diurnal cycle, resulting in a decrease in the temperature amplitude wave. Moreover, potential temperature profiles were reduced during daytime and increased at lower levels during the night, accordingly affecting the mixing height. The calculated results were compared with available measurements from the Mediterranean Campaign of Photochemical Evolution (MEDCAPHOT-TRACE) experimental campaign (Ziomas, 1998).

CENTRAL EUROPE AND UK

Between May and December 1994, Tumanov et al (1999) took measurements in Bucharest, Romania, using three ground-level air temperature measurements in a few representative points. Intensities were analysed in comparison with synoptic mesoscale patterns. The authors found that the intensity of the heat island is around 3.5°C and strongly increases immediately after sunset, keeping at a high and almost constant value all night, while it is at its minimum during the daytime. It was found that a strong dependence exists between the intensity of the heat island and weather patterns. The highest difference between urban and rural temperatures occurs in anticyclone situations, while the lowest is in frontal situations.

Pongracz et al (2006) studied surface heat island intensity in various Hungarian cities using satellite data. Average values of surface temperature differences between urban and rural areas range between 1°C and 6°C. Also in Hungary, Unger (1996) and Bottyan and Unger (2003) examined the influence of urban and meteorological factors on the intensity of heat islands in the medium-sized city of Szeged. They used mobile and stationary measurements under different weather conditions in February–March 2000 and April–October 2002. It was found that heat island intensity in the centre of the city is close to 2.1°C during the heating season and 3.1°C during the non-heating season (comparison refers to minimum daily temperatures). In parallel, it was found that there is a strong relationship between urban thermal excess and distance, as well as built-up ratio. Unger (1996) reports that the magnitude of urban effect on minimum temperature is larger in the case of anticyclonic weather conditions, and smaller during cyclonic weather conditions. By contrast, Bottyan and Unger (2003) report that meteorological conditions do not have any significant effect on UHI intensity at the time of its maximum development.

Szegedi and Kircsi (2003) studied heat islands in Debrecen, the second city of Hungary, with a population of 220,000. Measurements were carried out at ten-day intervals under various synoptic weather conditions using a mobile station during April 2002 and March 2003. Data were collected after sunset hours. It is reported that the mean maximum UHI intensity is 2.3°C, with an absolute maximum of 5.8°C. In the non-heating season stronger heat islands were detected than in the heating season. Similar results are also reported by Bottyan et al (2005), as well as by Szegedi (2006) for the city of Debrecen.

Szymanowski (2005) studied the variability in the air temperature field of the Wroclaw urban area in Poland during thermal advection, and its interaction with a UHI. Data from six urban and rural stations and for the period 1997–2000 were used. Mobile stations were also used. It was found that the temperature differences in the city during periods of advection are of short duration, only a few hours at most, but are sometimes large, reaching an intensity of 5–6°C and even as large as 9°C. It was also found that the thermal influence of advection is often greater than that due to urban factors. The intensity of the heat island during the day was 5–6°C, while during the night it was 8–9°C.

Klysik and Fortuniak (1999) analysed a series of synchronic meteorological measurements over a period of three years in the city of Lodz, Poland. Measurements from an urban and rural station were used and it was found that in special cases UHI intensity may reach more than 10°C, and sometimes may be close to 12°C. Two types of

UHI were observed: multi-cellular in windless and cloudless weather, and simple general in winds of 2–4m/s (greater wind velocity dissipates the urban heat surplus). The greatest differences occur during summer nights when skies are clear.

Blazejczyk et al (2006) report the amplitude of urban heat island intensity in various Polish cities. In Warsaw mean temperature heat island intensity is close to 1°C, while the corresponding values for the minimum and maximum temperatures are 3.1°C and 0.1°C. For Poznan, the heat island intensity corresponding to the mean, minimum and maximum daily temperatures are 2.1°C, 2.2°C and 3.8°C, respectively. For Bydgoszcz, the corresponding values are 0.1°C, 1.1°C and 1.6°C, while for Glucholazy the values are 0.8°C, 1.9°C and 3.5°C.

Beranova and Huth (2005) studied long-term trends in the relative intensity of the Prague UHI for the period 1961–1990. Comparisons are based on daily minimum temperatures at an urban station located in the city centre and three rural stations. It is reported that the intensity of the heat island increases in all seasons as well as annually. The increase is found to be significant in all seasons except for winter, when the trend is the weakest. The annual trend is close to 1.2°C per 100 years, while the corresponding trend for summer is close to 1.5°C per 100 years.

Dupont et al (1999) compared urban and rural temperatures in Paris using measurements collected during the ECLAP experiment performed during the winter of 1995. One urban site and one rural site were instrumented with sodars, lidars and surface measurements. Additional radiosondes, 100m/s masts and Eiffel Tower data were also collected. It was found that the heat island varies between 0°C and 6°C, with the maximum intensity presented at 08.00 hours.

Cristen and Vogtt (2004) present results from an experimental network of seven energy balance stations in and around Basel, Switzerland, during 2001 and 2002. The intensity of the nocturnal heat island was measured as 3°C and was observed just after sunset, with continuously decreasing values during the night.

Bohm (1998) reports an analysis of six urban and three rural temperature stations in Vienna, Austria. It was found that the urban excess temperatures vary from site to site, from 0.2°C in suburban areas up to 1.6°C in densely built-up areas. It is also reported that the urban effect in Vienna is more strongly influenced by the local surroundings of the site than by the city as a whole.

Lazar and Podesser (1999) report the results of extensive urban climate analysis in Gratz, Austria, completed in 1995. They report that the variation of thermal differences in the mean monthly temperatures is 2.2°C between the city centre UHI and the suburban areas. With mean everyday minima as the basis for determining the structure of the UHI, the difference between the city centre and suburban areas was 4.3°C in January. It is also reported that the differences are mainly a result of conditions at times of snow cover and human heat input in the city centre. In summer there are small differences of about 2°C, mainly as a result of nightly cloudiness after rain showers and thunderstorms.

Watkins et al (2002a) report the results of temperature measurements taken in London, UK, in summer 1999. The intensity of the London heat island was assessed using a radial grid of 68 stations recording simultaneous hourly air temperatures. The UHI was found to be predominantly a nocturnal phenomenon and its intensity reaches 7°C on occasion. It was found that the thermal centre is in the City of London, which is characterized by tall buildings and high anthropogenic heat emissions.

NORTHERN EUROPE

Shahoedanova et al (1997) studied the thermal climate in Moscow in 1990. They report an analysis of the air temperatures near the surface for a number of locations representing different land-use types. It was found that the urban/rural temperature differences ranged between 1°C and 3°C, with an absolute maximum of 9.8°C. Maxima are measured at night and the highest intensities of the heat island were observed between May and August.

Eliasson and Holmer (1990) studied UHI circulation in Gothenburg, Sweden. They found that heat island intensity during the night is at least 2.5°C. In a further study, Eliasson (1996) used stationary and mobile stations between 1989 and 1991 to investigate the climate in the city of Gothenburg. Nocturnal temperature distribution was analysed in relation to differences in street geometry and land use. It is shown that air temperature differences at night are small within the central city area, while the difference in the greatest mean air temperatures between the urban canyon and the open area is less than 3°C. It was also found that average urban–rural air temperature differences range from 3.5 to 6.0°C.

Hara and Autio (2006) report data on heat island intensity in the high latitude city of Oulu in Central Finland. Hourly temperatures were measured at three permanent stations between 1996 and 1998 using automatic data loggers. It is reported that heat island intensity during the winter period is 3.4°C.

SOME CONCLUSIONS

In the Mediterranean zone, most of the studies (Rome, Lisbon, Aveiro, Madrid, Granada, Turkish cities) concentrated on heat island intensity during the night period. Heat island intensity varies between 2°C for Istanbul, 7.5°C for Aveiro, and 9.0°C for the medium-sized Turkish cities. Apparently UHIs bear no relation to city size. In all the above cities, except Istanbul where data are not available, higher UHI intensities correspond to low wind speeds. For Lisbon, weather types with northerly winds are associated with the highest air temperatures in the downtown area, partly because of a shelter effect. Concerning the impact of cloud cover, studies for Aveiro, Madrid and Granada, show that the intensity of heat islands is at its maximum when the sky is totally clear. For Aveiro and Madrid, classification of UHI values on the basis of different types of weather shows that the highest values correspond to anticyclonic periods.

Maximum heat island intensities are presented during the daytime in Athens and Parma. High intensities were measured in Athens, where the maximum is found during the summer period. Higher UHI intensities were found during the summer period in Rome, Madrid and Parma. By contrast, maximum UHI intensity in Lisbon, based on minimum daily temperature, was found during the winter period.

Studies in central Europe and the UK concentrated on the analysis of heat islands during the night-time period. Data from all cities show that maximum UHI values developed at night. The maximum UHI intensity was reported for the city of Lodz (12°C), and the minimum for Vienna (1.6°C). Like the Mediterranean cities, UHIs in central Europe and the UK do not show any relation to the city size. Concerning the impact of the wind and cloud cover, data from Bucharest, Szeged, Debrecen, Wroclaw and Lodz, show that UHIs reach their greatest intensity in windless and cloudless conditions. For Prague, Debrecen, Lodz and Szeged, classification of UHI values on the basis of different types of weather shows that the highest values correspond to anticyclonic periods of weather. Heat islands are most developed during spring and summer in Bucharest, Debrecen and Wroclaw.

Urban heat island research in northern Europe, particular in Moscow and Gothenburg, shows that the highest intensities are observed during the night. The UHI for Moscow is around 9.8°C, and 6°C for Gothenburg. For Moscow, the highest intensities of the heat island were observed between May and August, while synoptic analysis of the period between May and August confirmed that low wind speed and clear skies associated with anticyclones generate strong heat islands.

THE ENERGY IMPACT OF HEAT ISLANDS

Increased urban temperatures have an important impact on the energy consumption of buildings, especially during the summer period. It is evident that higher urban temperatures increase electricity demand for cooling, and so the production of CO_2 and other pollutants also increases. By contrast, higher temperatures may reduce the heating load of buildings during the winter period.

In Europe, the impact of heat islands on the cooling and heating load of buildings has been extensively studied, as have the impact of heat islands on natural and night ventilation techniques in Athens and London.

In Athens, Hassid et al (2000) and Santamouris et al (2001) used multi-year data from a network of urban and rural stations and they calculated the spatial distribution cooling loads of buildings in the city. The authors found that the cooling load at the centre of the city is about double that of the surrounding region. The maximum cooling load corresponds to the very central area of Athens and especially to a station very close to a busy road. Minimum values were calculated in the south-east Athens region, a mean density residential area close to the Hemetus forest. Much higher cooling loads were calculated for the western Athens region, an area characterized by high density plots, lack of green spaces, important industrial activity and more traffic than the eastern Athens region. The difference between energy consumption for cooling in the larger municipality of western Athens (Aigaleo) and for cooling buildings situated at one specific reference station is equal to 180GW per hour.

Apart from increased energy loads for cooling of buildings, high ambient temperatures increase peak electricity loads and put local utilities under considerable strain. Thus, knowledge of the possible increase in peak electricity loads due to higher urban ambient temperatures is very important. For Athens, much higher peak cooling loads have been calculated for the central Athens area. For a set point temperature equal to 26°C, the highest peak load of the reference building was calculated as 27.5kW, while the minimum was 13.7kW. Thus, the impact of higher urban temperatures is extremely important and almost doubles the peak cooling load of the reference building.

When the set point temperature increases to 28°C, much higher differences have been calculated. The maximum cooling load was found to be 23.5kW while the minimum was 7.3kW. Thus, while the maximum peak cooling load is reduced by about 4.3kW, the minimum peak load is reduced by 6.4kW. The results indicate clearly that for the central Athens area the peak cooling load is mainly due to persistently very high ambient temperatures and is not sensitive to the change of the set point temperature. By contrast, the calculated minimum peak cooling load changes dramatically as the set point temperature increases. This is mainly due to the minimization of the load induced by the indoor–outdoor temperature differences as the set point increases. In this case the calculated peak cooling load is mainly due to solar radiation and other sources.

For western Athens the peak cooling power requirement estimated using the reference station data, is half that based on the data collected in the western Athens area. For the year 1997 the difference was 167MW, while for 1998 the difference was only 98MW. The corresponding estimates based on the Athens meteorological year are 53MW for 1997 and 146MW for 1998. If a diversity factor of 0.8 is assumed and a marginal price for an additional kW in the grid is equal to €1000, this gives an investment of approximately €93 millions in generation and distribution equipment, based on the average of the four figures mentioned above.

High ambient temperatures determine the efficiency of conventional air-conditioners. The coefficient of performance (COP) is directly affected by relative humidity and ambient temperature, and thus it is of great interest to those researching the possible decrease of the COP due to the heat island effect.

For Athens, by using hourly temperature and humidity data from all stations in and around the city for the summer of 1996, the spatial distribution of the minimum COP value of a conventional air-conditioning system was determined. It was found that the absolute minimum COP values correspond to the very central area of Athens (close to 75 per cent) because of the high ambient temperatures, as well as for the coastal area because of the high humidity. The highest minimum COP value was calculated at 102 per cent for the south-east area of Athens. Results show clearly that in addition to high cooling loads and peak electricity problems, heat islands significantly reduce (to about 25 per cent), the efficiency of air-conditioning systems and thus may oblige designers to increase the size of the installed systems, intensifying peak electricity problems and energy consumption for cooling purposes.

Increased urban temperatures because of the heat island effect may have an important impact on the heating load of urban buildings. Calculations of the spatial variability of the heating load of a reference building were performed across Athens. All stations were grouped into three clusters: stations located in the very central area, suburban stations, and those in urban parks and green areas. It was found that the heating load in the central Athens region was 3.7kWh/m^2/month, while the corresponding load of the suburban stations was 5.1kWh/m^2/month, and the mean load of the stations located in green areas was 7.3kWh/m^2/month. As shown, increased urban temperatures decrease the heating load of urban buildings to about 30 per cent while the maximum difference between suburban and urban stations was close to 55 per cent.

The direct and indirect environmental impact of the heat island effect in Athens has been estimated by Santamouris et al (2006). This was achieved through the estimation of the additional ecological footprint caused by the urban heat island phenomenon over the city. The first step of performing an ecological footprint estimation is made by calculating the increase in cooling demand caused by the heat island over the whole city and then by translating the energy into an environmental cost. Two years' annual experimental data from many urban stations were used. The results show that the ecological footprint because of the heat island ranges 1.5 to 2 times the city's official area; this area would have to be reserved every year to compensate for the additional CO_2 emissions caused by the heat island. The maximum potential ecological footprint caused if all buildings were air-conditioned would be almost 110,000 hectares.

Measured data from almost 80 surface stations collected in London were used by Watkins et al (2002b) as input to a thermal simulation model to assess the heating and cooling loads of a standard air-conditioned office building positioned at different locations in London. It was found that the energy demand for cooling always exceeds the need for heat, wherever the building is located. The urban cooling load is up to 25 per cent higher over a year, and the annual heating load is reduced by 22 per cent. Minimum CO_2 is emitted at a rural location, but the net rate of increase of CO_2 with increase in temperature is found to be 2.8 per cent per °C.

In another study by Kolokotroni et al (2006), data from the London heat island experiment were used as input to a thermal simulation model to assess the heating and cooling loads of a standard air-conditioned office building positioned at 24 different locations within the heat island. Eight urban sites were chosen to represent the range of sites from rural to central urban areas. It was found that the urban cooling load is up to 25 per cent higher than the rural load over a year, and the annual heating load is reduced by 22 per cent.

Caouris et al (2005) took ambient temperature measurements in the coastal city of Patras, Greece, using ten temperature sensors located in urban, rural and suburban areas. They report that the heating degree days (HDDs), for a base temperature of 14°C, in the urban areas is 39.1 per cent and 33.7 per cent less than the corresponding HDD for the rural and suburban areas. When heating degree hours (HDHs) are calculated for a base temperature of 18°C, the corresponding decreases of the HDH in the urban area are 22.3 per cent and 16.9 per cent, respectively.

Taesler et al (2006) assessed the impact of urban temperatures during winter in urban Sweden on the heating load of buildings. They report that increased urban temperatures may result in a decrease of the annual load by 10–15 per cent.

IMPACT ON NATURAL AND NIGHT VENTILATION POTENTIAL

The cooling potential of natural ventilation of buildings located in urban canyons is seriously reduced because of significant decrease in wind velocity inside the canyons and the corresponding increase in ambient temperature.

For Athens, in order to evaluate the natural ventilation potential of urban buildings, as well as the possible decrease in their cooling potential because of the canyon phenomenon, simulations of the airflow processes were carried out in ten canyons where wind speed and temperature data were collected (Geros et al, 2004). Two configurations were considered: a single-sided pattern and a cross-ventilation pattern.

Two types of simulation were performed for each configuration. The first was based on wind and temperature data taken inside each canyon, while the second was based on the undisturbed temperature and wind speed measured over the buildings. Comparison of both simulation results allowed the assessment of the decrease in the cooling potential of natural ventilation in urban canyons.

It was found that during the daytime, when the ambient wind speed is considerably higher than wind speeds inside the canyon and when inertia phenomena dominate the gravitational forces, the natural ventilation potential in single-sided and cross-ventilation configurations is seriously decreased. In practice this happens when the ambient wind speed is higher than 4m/s. For single-sided ventilation configurations the airflow is reduced by up to five times, while in cross-ventilation configurations the flow is sometimes reduced by up to ten times. During the daytime but when the ambient wind speed is lower than 3–4m/s, gravitational forces dominate the airflow processes. In this case the difference in wind speed inside and outside the canyon does not play any important role, especially in single-sided configurations. Finally, during the night time ambient wind speeds are much decreased and are comparable to wind speeds inside the canyon. In this case the airflow calculated for inside and outside the canyon is almost the same.

The specific performance of night ventilation techniques in urban environments suffering from heat island problems was studied by Geros et al (2004). The work determined the impact of the urban environment on night ventilation energy performance by studying a typical room located in the centre of Athens under air-conditioned and free-floating operation, considering single-sided and cross-ventilation. The influence of the urban microclimate on the efficiency of night ventilation was examined by considering the typical zone inside the canyons and under undisturbed conditions. The comparison of the results allowed the impact of the urban environment on the effectiveness of night ventilation techniques to be explored, showing that due to the increase in air temperature and the decrease in wind velocity inside the canyons, the efficiency of the studied techniques is significantly reduced when compared with the undisturbed conditions dominating outside the urban canyons. In conclusion, when the ventilation is single-sided, the relative increase in the cooling load between the two positions of the typical zone varies from 6 per cent to 89 per cent. For cross-ventilation, this difference varies between 18 per cent and 72 per cent. Consequently the use of climatic data measured outside the canyon overestimates, sometimes significantly, the energy performance of night ventilation.

The impact of the London UHI on the summer cooling demand of buildings and their night ventilation strategies was studied by Kolokotroni et al (2006). A reference and optimized office module is defined and used for analysis. The results are presented for two representative weeks using the available climatic data, one for extreme hot weather and one for typical hot weather, within the London heat island and the rural reference site. A comparison of the building types based in the same location suggests that during the typical hot week a rural reference office has 84 per cent of the energy demand for cooling compared to a similar urban office. A rural optimized office using night ventilation techniques would not need any artificial cooling and would be able to maintain temperatures below 24°C. An urban optimized office, also using night ventilation techniques, would not be able to achieve this. A rural optimized office would need 42 per cent of the cooling energy required for an optimized urban office. A comparison of the optimized to reference office modules suggests that an urban optimized office reduces the cooling demand to 10 per cent of the urban reference office.

IMPACT OF HEAT ISLANDS ON POLLUTION

Increased urbanization, energy consumption and growing pollutant emissions, associated with the presence of UHIs, have a strong impact on urban pollution (Crutzen, 2004). Crutzen argues that it is 'important to explore the consequences of combined urban heat and pollution island effects for meso-scale dynamics and chemistry'.

To this end, Sarrat et al (2006) studied the impact of urban heat islands on pollution during a summertime anticyclonic episode associated with high photochemical pollution in the Paris region in France. The study concluded that both nocturnal and diurnal urban effects have an important impact on primary and secondary regional pollutants and more specifically the levels of ozone and nitrogen oxides. The spatial distribution and presence of pollutants are significantly modified by the urbanized area mainly due to enhanced turbulence.

Stathopoulou et al (2006) studied the impact of increased ambient temperatures on tropospheric ozone concentration in the urban area of Athens. Air temperature and ozone concentration data from several experimental stations in the Greater Athens area were collected and used. A strong correlation between urban temperatures and ozone concentrations was found. In parallel, it is observed that the fluctuation in ozone concentration follows relative ambient temperature variations, especially during the daytime when the highest values of ozone are observed because of high levels of solar radiation and ozone precursors.

MODELLING AND PREDICTION OF HEAT ISLANDS

Much research is dedicated to the modelling and prediction of UHIs. According to Atkinson (2003), modelling follows three approaches: hardware modelling, physical modelling using the surface energy budget equation, and dynamical numerical modelling. Other non-deterministic approaches may also be used. Data-driven techniques based on intelligent algorithms, such as neural networks or statistical correlation methods, have been developed recently.

Although it is beyond the scope of this paper to present the state of the art of UHI modelling, some interesting approaches have recently been applied in Europe and are briefly outlined here.

Deterministic models follow three main approaches: modelling of urban canyon temperature regimes (Sánchez de la Flor and Alvarez Dominguez, 2004), modelling of the urban canopy layer (Bonacquisti et al, 2006), and dynamic numerical modelling (Atkinson, 2003).

Sánchez de la Flor and Alvarez Dominguez (2004) propose an urban canyon thermal model to predict the ambient temperature as well as the surface temperature of buildings, pavements and streets. The model is based on the simultaneous solution of the thermal balance equation for each surface and the ambient air in the canyon. Airflow phenomena are not calculated and the airflow pattern is an input to the model.

Bonacquisti et al (2006) develop an urban canopy prediction model. The model uses four equations: the energy balance at the surface of buildings, the energy balance at ground level, the sensible heat flux, and the latent heat flux. The model provides as output the skin temperature of buildings, air temperature and humidity within the canopy layer and hence the mean surface temperature and the air temperature at 2m above surface level. The model was applied to Rome during radiative summer and winter episodes and validated against observations.

A three-dimensional, non-hydrostatic, high-resolution numerical model is proposed and used by Atkinson (2003) to analyse UHI intensity in London. The urban area was represented by anomalies of albedo, anthropogenic heat flux, emissivity, roughness, length, sky view factor, surface resistance to evaporation, and thermal inertia. The model was validated against observations.

A data-driven prediction model based on linear regression techniques is presented by Bottyan and Unger (2003). The model predicts the mean maximum value of UHI intensity in both seasons as a function of basic parameters: the mean sky view factor, mean building height, the ratio of built-up surface, and the ratio of water surface. The authors claim that the linear regression model may be used to predict UHI in other cities with similar boundary conditions.

Intelligent techniques such as neural networks have been used to predict UHI intensity (Santamouris et al, 1999). A neural network approach to predict the spatial distribution of UHI intensity in Athens during the day and night periods is presented by Mihalakakou et al (2004). The authors prepared two neural network models, one for the day and the other for night-time. Both models predict the maximum temperature difference during the corresponding period compared to the reference station. For the night period the used inputs are the maximum daily values of the ambient air temperature (Tmax in °C), measured at each urban station, as well as night-time values of the ambient air temperature (Tref in °C), measured at the reference station, at the time when $DTmax_{night}$ is observed.

For the daytime period, the used inputs are: daytime values of ambient air temperature (Tair in °C), measured at each urban station at the time when the $DTmax_{day}$ is observed, maximum daily values of global solar radiation, and daytime values of reference air temperature (Tref in °C), measured at the reference station at the time when the $DTmax_{day}$ is observed. Comparison of the predicted against the measured values revealed strong agreement. The model may be used to predict the spatial distribution of the heat island in Athens as a function of climatological data measured by standard meteorological stations.

Mihalakakou et al (2002) propose a neural network model to predict maximum heat island intensity in Athens. Five inputs were used: synoptic atmospheric circulation, ambient air temperature measured at various stations, ambient air temperature at the reference station, maximum daily values of solar radiation, and mean daily wind speed. An excellent agreement between experimental and theoretical values was found.

RESEARCH ON MITIGATION TECHNIQUES

Several techniques have been proposed to mitigate heat islands of which two appear to be especially significant: the use of green spaces, and the use of appropriate materials in buildings and the urban fabric.

THE ROLE OF GREEN SPACES

The importance of green spaces to mitigate heat islands has been stressed by many European researchers (Harrison et al, 1995; Eliasson, 2000; Handley et al, 2003). Trees create a favourable thermal balance for people and enhance outdoor thermal comfort (Picot, 2003). Papadakis et al (2001) investigate the ability of trees to control solar radiation on vertical façades in Greece. They report that 70–85 per cent of incident radiation was intercepted by the trees, while the ambient temperature behind the shaded area was lower in comparison with those façades without trees.

Parks can reduce urban temperatures; temperature reductions depend on park size and the distance from buildings to the park. Gomez et al (1998) compared urban temperatures of green spaces and of other locations in the city of Valencia, Spain. They report that in the green areas, even in situations that were not favourable for forming heat islands, there was a drop of 2.5°C with respect to the city's maximum temperature. Von Stülpnagel (1987) and Von Stülpnagel et al (1990) report that daytime air temperatures in a large urban park in Berlin, Germany, were found to be over 2°C lower than surrounding built-up areas. According to the same studies, urban parks reduce air temperatures in the adjacent neighbourhood; however, this effect was limited to a relatively small zone extending only 200–400m from the margin of a large park on a calm day.

Measurements of the ambient temperature in and around an urban park were taken for a period of ten days in Athens during August 1998 (Santamouris, 2001). Measurements were taken simultaneously using five mobile stations moving from the centre of the park to its exterior space during the daytime period. The aim of the experiment was to investigate the temperature difference between the vegetated and the surrounding area, as well the gradient of temperature increase as a function of distance from the park. It is reported that:

- The temperature inside the park varied mainly as a function of shading and vegetative cover. The difference between the lower and higher temperatures was as high as 1.5°C.
- The maximum temperature difference between the park and the surrounding urban area during the day period was 3°C.
- There was no constant gradient of temperature increase with distance from the park. At all four entrances to the park, it has been found upon exiting there was an immediate temperature increase of about 1°C. This almost had the characteristics of a step function.
- The temperature around the park was mainly influenced by other factors than the presence of the park, such as the density of the buildings, the rate of anthropogenic heat released mainly by cars, the shading of the canyons and so on.

Santamouris (2001) also reveals the findings of research using multi-year temperature measurements taken in two urban parks in Athens. They show that both parks have almost 40 per cent less cooling degree hours than the other surrounding urban stations. The parks also present the lowest recorded absolute minimum temperatures of all the stations. However, during the daytime the absolute temperature inside the park was higher than at the suburban reference station. During the heating period, the urban parks had 15 per cent higher HDDs than the surrounding stations. During the summer season and around noon, both parks had temperatures 4–7°C lower than that of the surrounding stations, but 2–5°C higher than that of the suburban reference station. During summer nights both parks had temperatures 5–6°C lower temperature than the urban reference station and 2–3°C lower than the suburban reference station.

According to Handley et al (2003) parks need to have a size of at least one hectare to have a significant climatic effect and thus a dense network of public green spaces is necessary. Pauleit and Duhme (1995; 2000) stress that public green space cannot compensate for lack of vegetation within urban land uses.

Finally, concerning modelling of the thermal impact of green spaces in cities, Robitu et al (2006) develop a numerical model in quasi-steady state, which allows the evaluation of the impact of trees and water ponds on the urban thermal environment and the comfort of pedestrians. Dimoudi and Nikolopoulou (2003), also model the thermal impact of vegetation in the urban environment using computational fluid dynamics.

Planted roofs can contribute significantly to the mitigation of heat islands. Planted roofs present much lower temperatures than hard surfaces and decrease the ambient temperature through convection and evapotranspiration. Important research has been carried out in Europe to investigate the impact of planted roofs on the thermal performance of buildings, as well as on the urban environment. Eumorfopoulou and Aravantinos (1998) simulated various planted roof elements with different heights of plants and different drainage and they performed comparisons between a bare roof and a planted roof. They conclude that the planted roof contributes highly to the thermal protection of buildings, but does not replace the thermal insulation layer. Niachou et al (2001) report on extensive measurements of a planted roof in Greece and they concluded that it significantly reduced the cooling load of the building.

Various predictive models for green roofs have also been developed. Del Barrio (1998) proposes a simplified model and performed parametric sensitivity analyses to assess the cooling potential of green roofs in summer. It was found that green roofs do not act as cooling devices but as insulation devices, reducing the heat flux through the roof. Theodosiou (2003) proposes a thermal model to simulate planted roofs and validated the model using data from a real construction. Another predictive model, validated using real data, is proposed by Lazzarin et al (2005).

COOL MATERIALS FOR THE URBAN ENVIRONMENT
Heat island intensity can be significantly reduced by decreasing the thermal gains in the urban environment, and in particular the amount of solar radiation absorbed. This can be achieved by increasing the albedo of cities by using materials for buildings and the urban fabric that present high reflectivity to solar radiation. Such materials exist in nature and are widely used in warm climates. Also, artificial materials with high reflectivity to short-wave radiation and high emissivity values have been created. These 'cool materials' have become the subject of much research recently (Akbari et al, 1992; Bretz and Akbari, 1997). A full description of the actual state of the art on appropriate materials to mitigate heat islands is given by Santamouris (2001).

European research on cool and appropriate materials for the urban environment can be divided into four phases, each of which is described in more detail below:

- Phase one – study of the thermal and optical characteristics and performance of natural reflecting materials;
- Phase two – development and testing of white materials presenting high reflectivity in the visible spectrum and high emissivity values;
- Phase three – development and testing of white materials presenting high reflectivity in the visible spectrum and low emissivity values;
- Phase four – development and testing of coloured materials presenting a high reflectivity to the near infrared solar spectrum.

Natural reflecting materials

Doulos et al (2004) performed a comparative study aiming to investigate the suitability of materials used in outdoor urban spaces to contribute to lower ambient temperatures and reduce heat island effect. The study involved a total of 93 pavement materials commonly used outdoors and was performed during the summer period of 2001. The selected sample materials consisted of several different construction materials (concrete, asphalt, marble, granite, paving stone, stone, pebble and mosaic) of different surface colour materials (white, grey, black, red, brown and green), and of different surface texture materials (with smooth surfaces, rough surfaces and anaglyph surfaces with marks and designs).

All materials where exposed to solar radiation and their surface temperatures were measured using surface and infrared temperature sensors. As all materials presented an emissivity value close to 0.9, the results permitted an evaluation of the impact of different reflectivity values and the characteristics of their thermal performance. Colour, the surface texture and the construction material were all found to affect albedo.

As expected, the minimum values of the mean daily and the absolute maximum surface temperatures were observed for the white coloured tiles, while the maximum values were measured for the dark coloured tiles. The mean daily surface temperatures ranged between 29.7°C for the white marble tile, and 46.7°C for the asphalt tile. The absolute maximum temperatures ranged between 33.4°C and 54°C for the same two materials.

With white surfaces, the observed temperature differences were due to differences in surface textures. In general, the smooth surfaced materials presented lower surface temperatures than those with rough surfaces or anaglyph schematics. The authors conclude that most of the studied materials presented higher mean surface temperatures than the average ambient temperature. Only the white coloured tiles made of marble were cooler than the ambient air. The warmest (38.1°C) light coloured tile was made of pebble with white and green surface colour; the maximum temperature difference between its light coloured materials and the ambient air was close to 6.9°C. For dark coloured materials, the maximum temperature difference was observed with asphalt and was close to 15.5°C. The coldest dark coloured material was one made of stone presenting a mean surface temperature close to 41.4°C and a temperature difference with the ambient air close to 10.2°C.

It is also concluded that for black coloured surfaces the lowest temperatures were measured for those made of mosaic, concrete and marble. For white materials the lowest temperatures were for those made of mosaic, concrete, granite, pebble and marble. For grey surfaces all the materials presented low temperatures except for pebble and paving stone. For green materials the lowest temperatures were observed for mosaic and granite. Finally, brown materials made of mosaic and stone had the lowest temperatures.

White materials with high reflectivity and high or low emissivity

Synnefa et al (2006a) performed a comparative study of 14 types of white reflective coatings, aiming to investigate the thermal effect of these on lowering surface temperatures of buildings and other surfaces in the urban environment, and thus test their potential to lower ambient temperatures and reduce the heat island effect. All materials were tested between August and October 2004 on a 24-hour basis. The thermal as well as the optical characteristics of all materials were measured.

The materials tested were: a silver-grey aluminium pigmented acrylic coating, an acrylic ceramic coating, three acrylic elastomeric coatings, an alkyd chlorine rubber coating, an aluminium pigmented alkyd coating, an emulsion paint, an acryl-polymer emulsion paint, an acrylic latex, an aluminium pigmented coating, an acrylic insulating paint, a silver aluminium pigmented acrylic coating, an epoxy polyamide coating, a white acrylic paint, and finally an uncoated tile used as reference.

All materials were exposed to solar radiation. Surface temperatures, emissivity and spectral reflectance of every material were measured. As expected, the minimum values of the mean and the maximum daily surface temperature were observed for the white coatings, while the maximum corresponding values were recorded for the silver coloured coatings.

Measurements of the optical characteristics revealed that the spectral reflectance of silver paints that contain aluminium pigments increases with increasing wavelength and suddenly drops at around 800nm. White coatings with the higher reflectance appear to stay cooler during the day. Among the white coatings the worst thermal performance was observed for coatings whose reflectance curve stays under 80 per cent. The spectral reflectance curves of the silver coloured coatings are lower than the white coloured coatings.

Measurements have shown that during the daytime the thermal performance of the samples is mainly affected by their surface solar reflectance, while emissivity has a lower impact. During the night, emissivity becomes the predominant factor affecting the thermal performance of the tiles. It was found that the mean nocturnal surface temperature and the infrared emissivity value of the samples were highly correlated (with a correlation coefficient of 0.84). The analysis revealed that white coatings present lower temperatures during the day compared to the aluminium pigmented coatings. Although all the coatings studied are characterized by quite high solar reflectance, aluminium coatings stay warmer during the night due to their lower infrared emissivity.

Measurements also showed that during the daytime period the white coloured coatings in general had the ability to significantly reduce the surface temperature of the concrete tile on which they were applied. The maximum temperature decrease was observed for a white acrylic elastomeric coating, a white acryl-polymer emulsion paint, a white alkyd, chlorine rubber coating and a white acrylic ceramic coating, these reduced the mean concrete tile's surface temperature by 7.5°C, 5°C, 4.3°C, 4°C and 4°C, respectively. By contrast, the silver coloured coatings were found to increase the concrete tile's surface temperature. The maximum increase of temperature was observed for a silver aluminium pigmented coating, with a mean surface temperature about 6.3°C higher than the uncoated tile's surface temperature.

During the night, ten coatings presented surface temperatures lower than those of the uncoated concrete tile. The coolest coatings were the white acrylic ceramic coating, the white acrylic elastomeric coatings and the white acryl-polymer emulsion paint. These have higher infrared emittance values and reduced mean surface temperatures by an average of 2°C.

The impact of weathering and 'dirt pick-up' resistance was also evaluated. Weathering is caused by surface contamination and other alterations such as ultraviolet radiation, sudden temperature swings and moisture penetration. An important degradation in the thermal performance of several coatings, and mainly of the acrylic elastomeric coating, the white acrylic paint and the white acrylic latex, was observed. The most important change in thermal behaviour was observed for the acrylic elastomeric coating, which was the coolest coating during the daytime period of the first month of the experiment, and then became a lot warmer during the second and third months of the experimental period.

Comparison of the mean surface temperature of the various coatings against the mean ambient air temperature revealed that all the coatings were characterized by higher average surface temperatures than the average air temperature during the daytime period. The coolest coatings were coloured white. Among these, the acrylic ceramic coating, the alkyd, chlorine rubber coating and the acryl-polymer emulsion paint were warmer than the ambient air by only 2°C, while the acrylic insulating paint and the epoxy polyamide coating were an average of 6°C warmer than the ambient air. For the silver coloured materials the minimum temperature difference of 9.6°C was measured for the aluminium pigmented acrylic coating, while the maximum temperature difference of 11.7°C was recorded for the aluminium pigmented coating.

All coatings were compared against two cool materials: a white marble and a white mosaic tile. It was found that during daytime, the white acrylic ceramic coating, the two acrylic elastomeric coatings, the alkyd, chlorine rubber coating, the acryl-polymer emulsion paint, the acrylic latex and the acrylic paint presented lower mean surface temperatures than both the marble's and the mosaic's mean surface temperatures. The maximum difference in mean surface temperature was found to be between the acrylic ceramic coating and the marble tile (3.8°C) and between the acrylic elastomeric coating and acrylic emulsion paint and the mosaic tile (4.1°C). Differences between the mean maximum daily temperatures were even greater. The mean maximum surface temperature of the acrylic ceramic coating was found to be 5.3°C and 6.9°C lower than the maximum surface temperatures of the marble tile and the mosaic tile, respectively. During the night period the temperature differences were not significant.

Coloured materials with high reflectivity to the near infrared

Synnefa and colleagues (Synnefa and Santamouris, 2006; Synnefa et al, 2006b) reported measurements of the solar spectral properties as well as of thermal performance of ten prototype cool coloured coatings developed at the National and Kapodistrian University of Athens. Specifically, orange, light blue, blue, green, two black, anthracite, brown, chocolate brown and light brown coatings were developed and tested. The coatings were developed using near infrared reflective colour pigments and were compared to colour-matched, conventionally pigmented coatings. The special infrared-reflecting colour pigments used in the prototype cool coloured coatings are complex inorganic pigments that exhibit high reflectance in the near infrared portion of the electromagnetic spectrum.

The coatings were developed for use in the urban environment to counter the heat island effect. The spectral reflectance and the infrared emissivity were measured and the solar reflectance of the samples was calculated. The surface temperatures of the coatings (applied on concrete tiles) were monitored using surface temperature sensors and a data logging system, with measurements taken on a 24-hour basis from August to December 2005.

It was found that for the first and the second month of the experimental period, mean daily surface temperatures ranged between 36.4°C and 31.6°C (for cool orange) and 49.4°C and 40.9°C (for black (2) standard), respectively. As expected, the coatings with the higher values of solar reflectance demonstrated the lower surface temperatures. During the day, all the cool coloured coatings had surface temperatures lower than the standard colour-matched coatings. The best performing cool coatings were black (2), chocolate brown, blue and anthracite, with mean daily surface temperatures differing from the standard colour-matched coatings by 5.2°C, 4.7°C, 4.7°C and 2.8°C, respectively, for the month of August. The highest temperature difference was observed between cool and standard black (2) and was equal to 10.2°C. The lowest temperature difference was observed between cool and standard green (2) and was equal to 1.6°C.

During the night when there is no solar radiation, the surface temperatures of the samples were quite uniform owing to all the coatings having an emissivity of around 0.88. However, cool coloured coatings remained cooler (by 0.1–1.6°C) than the standard colour-matched coatings, probably because they had absorbed smaller amounts of solar radiation during the day.

The temperature difference between the cool and standard coatings decreased from August to December as the monthly average daily global solar radiation decreased too, and the impact of the infrared-reflecting pigments in the coatings became less evident. The mean maximum temperature difference between the standard and cool black (2) coating was 6.5°C during August and dropped to 0.5°C for the month of December.

In order to evaluate the impact of a large-scale increase of urban albedo and the possible decrease of the surface and ambient temperatures during the summer period in Athens, numerical simulations were performed by the 'urbanized' version of the non-hydrostatic PSU/NCAR Mesoscale Model (MM5, version V3-6-1) (Synnefa et al, 2006c). The whole process was supplemented by detailed information on land-use cover derived from satellite image analysis (with a spatial resolution of 30m). The grid was equal to 0.666 x 0.666km². Simulations were performed for a typical summer day (15 August) and two scenarios were investigated. The base case scenario considered a building roof albedo equal to 0.18, while the alternative scenario considered a roof albedo of a building close to 0.63. The latter can be achieved using the cool coloured materials described above. It was found that the expected decrease of the ambient temperature at 2m height in the city centre at 15.00 hours local time varied between 0.8°C and 1.6°C.

CONCLUSIONS

This paper presented the findings of recent research on urban ambient air heat islands in Europe. Heat islands are an important climatic phenomenon in Europe and are present in all geographic zones of the continent. The intensity of the phenomenon is related to the main parameters identified by research as wind, cloud cover and the existence of cyclonic or anticyclonic conditions. Heat islands have an important impact on energy demand during the summer period and studies in Athens and London have shown that UHIs are associated with much higher cooling loads for buildings, increases in peak electricity demand, and decreases in the performance of air-conditioners and in the cooling potential of natural and night ventilation techniques. UHIs also have an important global environmental impact because they increase the ecological footprint of cities.

Important deterministic and data-driven models have been developed to predict UHIs. The use of intelligent techniques like neural networks permit the prediction of heat island intensity in a city as a function of the main climatic parameters. In parallel, important research has been carried out aiming to help mitigate heat island intensity. Research has focused on the understanding of the impact of urban green areas, as well as on the development of appropriate materials for the urban environment. In particular, white and cool coloured materials have been developed and tested. It has been found that cool materials present much lower surface temperatures than common materials and this can contribute significantly to lowering urban ambient temperatures and mitigate the effects of heat islands in Europe.

AUTHOR CONTACT DETAILS

Mat Santamouris, Group of Building Environmental Research, Physics Department, University of Athens, Greece
Email: msantam@phys.uoa.gr

REFERENCES

Akbari, H., Davis, S., Dorsano, S., Huang, J. and Winett, S. (1992) *Cooling Our Communities: A Guidebook on Tree Planting and Light Colored Surfacing*, US Environmental Protection Agency, Office of Policy Analysis, Climate Change Division, San Franscisco, CA

Akbari, H., Levinson, R., Miller, W. and Berdahl, P. (2005) 'Cool colored roofs to save energy and improve air quality', *Proceedings of International Conference on Passive and Low Energy Cooling for the Built Environment*, Santorini, Greece

Alcoforado, M. J. (2002) *O Clima da Região de Lisboa: Contrastes e Ritmos Térmicos*, PhD thesis, Memórias do Centro de Estudos Geográficos, Lisbon, CEG 15

Alcoforado, M. J. and Andrade, H. (2006) 'Nocturnal urban heat island in Lisbon (Portugal): Main features and modelling attempts', *Theoretical Applied Climatology*, vol 84, pp151–159

Arnfield, A. J. (2003) 'Two decades of urban climate research: A review of turbulence, exchanges of energy and water, and the urban heat island', *International Journal of Climatology*, vol 23, pp1–26

Atkinson, B. W. (2003) 'Numerical modelling of urban heat-island intensity', *Boundary-Layer Meteorology*, vol 109, no 3, pp285–310

Bacci, P. and Maugeri, M. (1992) 'The urban heat island of Milan', *Nuovo Cimento*, vol 15, C 4, pp417–424

Balkestahl, L., Monteiro, A., Gois, J., Taesler, R. and Quenoi, H. (2006) The influence of weather types on the urban heat island's magnitude and patterns at Paranhos, Oporto – a case study from November 2003 to January 2005, *Proceedings of the International Conference on Urban Climatology*, Gotemborg, 2006

Beranova, R.. and Huth, R. (2005) 'Long-term changes in the heat island of Prague under different synoptic conditions', *Theoretical Applied Climatology*, vol 82, pp113–118

Blazejczyk, K., Bakowska, M. and Wieclaw, M. (2006) 'Urban heat island in large and small cities', *Proceedings of 6th International Conference on Urban Climate*, Gothenburg, Sweden

Bohm, R. (1998) 'Urban bias in temperature time series: A case study for the city of Vienna, Austria', *Climatic Change*, vol 38, pp113–128

Bonacquisti, V., Casale, G. R., Palmieri, S. and Siani, A. M. (2006) 'A canopy layer model and its application to Rome', *Science of the Total Environment*, vol 364, pp1–13

Bottyan, Z. and Unger, J. (2003) 'A multiple linear statistical model for estimating the mean maximum urban heat island', *Theoretical Applied Climatology*, vol 75, pp233–243

Bottyan, Z., Kircsi, A., Szegedi, S. and Unger, J. (2005) 'The relationship between built-up areas and the spatial development of the mean maximum urban heat island in Debrecen, Hungary', *International Journal of Climatology*, vol 25, pp405–418

Bretz, S. E. and Akbari, H. (1997) 'Long-term performance of high albedo roof coatings', *Energy and Buildings*, vol 25, pp159–167

Caouris, Y. G. Giannopoulos, A. and Santamouris, M. (2005) 'Heating demands differences between central and surrounding areas in the coastal town of Patras', *Proceedings of the International Conference Passive and Low Energy Cooling*, Santorini, Greece

Colacino, M. and Lavagnini, A. (1982) 'Evidence of the urban heat island in Rome by climatological analyses', *ArchIVES. Meteorology. Geophysics. Bioklimatology.*, Series B, vol 31, pp87–97

Cristen, P. and Vogtt, W. (2004) 'Energy and radiation balance of a central European city', *International Journal of Climatology*, vol 24, pp1395–1421

Crutzen, P. (2004) 'New directions: The growing urban heat and pollution "island" effect – impact on chemistry and climate', *Atmospheric Environment*, vol 38, pp3539–3540

Dandou, A., Tombrou, M., Akylas, E., Soulakellis, N. and Bossioli, E. (2005) 'Development and evaluation of an urban parameterization scheme in the Penn State/NCAR Mesoscale Model (MM5)', *Journal of Geophysical Research– Atmospheres*, vol 110

Del Barrio, E. P. (1998) 'Analysis of the green roofs cooling potential in buildings', *Energy and Buildings*, vol 27, no 2 pp179–193

Dimoudi, A. and Nikolopoulou, M. (2003) 'Vegetation in the urban environment: Microclimate analysis and benefits', *Energy and Buildings*, vol 35, 69–76

Doulos, L., Santamouris, M. and Livada, I. (2004) 'Passive cooling of outdoor urban spaces: The role of materials', *Solar Energy*, vol 77, pp231–249

Dupont, L., Menut, B., Carissimo, J., Pelon, P. and G. Flamant (1999) 'Comparison between the atmospheric boundary layer in Paris and its rural suburbs during the ECLAP experiment', *Atmospheric Environment*, vol 33, pp979–994

Eliasson, I. (1996) 'Urban nocturnal temperatures, street geometry and land use', *Atmospheric Environment*, vol 30, no 3, pp379–392

Eliasson, I. (2000) 'The use of climate knowledge in urban planning', *Landscape and Urban Planning*, vol 48, pp31–44

Eliasson, I. and Holmer, B. (1990) 'Urban heat island circulation in Goteborg, Sweden', *Theoretical Applied Climatology*, vol 42, pp187–196

Eumorfopoulou, E. and Aravantinos, D. (1998) 'The contribution of a planted roof to the thermal protection of buildings in Greece', *Energy and Buildings*, vol 27, pp20–36

Geros, V., Santamouris, M., Karatasou, S., Tsangrassoulis, A. and Papanikolaou, N. (2004) 'On the cooling potential of night ventilation techniques in the urban environment', *Energy and Buildings*, vol 37, pp243–257

Gomez, F., Gaja, E. and Reig, A. (1998) 'Vegetation and climatic changes in a city', *Ecological Engineering*, vol 10, pp355–360

Handley, J., Pavleit, S., Slinn, P., Barber, A., Baker, M., Jones, C. and Lindley, S. (2003) *Accessible Natural Green Space Standards in Towns and Cities: A Review and Toolkit for their Implementation*, English Nature Research Reports, Peterborough, UK

Hara, Y. and Autio, J. (2006) 'Heat island intensity at high latitude city: An example from Oulu, Central Finland', *Proceedings of 6th International Conference on Urban Climate*, Gothenburg, Sweden

Harrison, C., Burgess, J., Millward, A. and Dawe, G. (1995) *Accessible Natural Greenspace in Towns and Cities: A Review of Appropriate Size and Distance Criteria*, English Nature Research Reports, no 153, English Nature, Peterborough

Hassid, S., Santamouris, M., Papanikolaou, P., Linardi, A. and Klitsikas, N. (2000) 'The effect of the heat island on air conditioning load', *Energy and Buildings*, vol 32, no 2, pp131–141

Karaca, M., Tayant, M. and Toros, H. (1995) 'Effects of urbanisation on climate of Istanbul and Ankara', *Atmospheric Environment*, vol 29, no 23, pp3411–3421

Klysik, K. and Fortuniak, K. (1999) 'Temporal and spatial characteristics of the urban heat island of Lodz, Poland', *Atmospheric Environment*, vol 33, pp3885–3895

Kolokotroni, M., Giannitsaris, I. and Watkins, R. (2006) 'The effect of the London urban heat island on building summer cooling demand and night ventilation strategies', *Solar Energy*, vol 80, no 4, pp383–392

Kolokotroni, M., Zhang, Y. and Watkins, R. (2007) 'The London heat island and building cooling design', *Solar Energy*, vol 81, no 1, pp102–110

Lazar, R. and Podesser, A. (1999) 'An urban climate analysis of Graz and its significance for urban planning in the tributary valleys east of Graz (Austria)', *Atmospheric Environment*, vol 33, pp4195–4209

Lazzarin, R., Castellotti, F. and Busato, F. (2005) 'Experimental measurements and numerical modelling of a green roof', *Energy and Buildings*, vol 37, pp1260–1267

Livada, I., Santamouris, M., Niachou, K., Papanikolaou, N. and Mihalakakou, G. (2002) 'Determination of places in the great Athens area where the heat island effect is observed', *Theoretical Applied Climatology*, vol 71, pp219–230

Mihalakakou, P., Flocas, H., Santamouris, M. and Helmis, C. (2002) 'Application of neural networks to the simulation of the heat island over Athens, Greece, using synoptic types as a predictor', *Journal of Applied Meteorology*, vol 41, no 5, pp519–527

Mihalakakou, P., Santamouris, M., Papanikolaou, N. and Cartalis, C. (2004) 'Simulation of the urban heat island phenomenon in Mediterranean climates', *Journal of Pure and Applied Geophysics*, vol 161, pp342–367

Montavez, J., Rodriguez, A. and Jimenez, J. (2000) 'A study of the urban heat island of Granada', *International Journal of Climatology*, vol 20, pp899–911

Moreno-Garcia, M. C. (1994) 'Intensity and form of the urban heat island in Barcelona', *International Journal of Climatology*, vol 14, pp705–710

Niachou, A., Papakonstantinou, K., Santamouris, M., Tsagrasoulis, A. and Mihalakakou, P. (2001) 'Analysis of a green roof thermal properties and investigation of its energy performance', *Energy and Buildings*, vol 33, pp719–729

Oke, T. R. (1982) 'The energetic basis of the urban heat island', *Quarterly Journal of the Royal Meteorological Society*, vol 108, pp1–24

Papadakis, G., Tsamis, P. and Kyritsis, S. (2001) 'An experimental investigation of the effect of shading with plants for solar control of buildings', *Energy and Buildings*, vol 33, pp831–836

Pauleit, S. and Duhme, F. (1995) 'Developing quantitative targets for urban environmental planning', *Land Contamination and Reclamation*, vol 3, no 2, pp64–66

Pauleit, S. and Duhme, F. (2000) 'Assessing the environmental performance of land cover types for urban planning', *Landscape and Urban Planning*, vol 52, pp1–20

Petralli, M., Morabito, M., Bartolini, G., Torrigiani, T., Cecchi, L. and Orlandini, S. (2006) 'Air temperature distribution at pedestrian level in Florence, Italy, through fixed and mobile sensor measurements', *Proceedings of 6th International Conference on Urban Climate*, Gothenburg, Sweden

Picot, X. (2003) 'Thermal comfort in urban spaces: Impact of vegetation growth. Case study: Piazza della Schienza, Milan, Italy', *Energy and Buildings*, vol 36, pp329–334

Pinho, O. S and Manso Orgaz, M. D. (2000) 'The urban heat island in a small city in coastal Portugal', *International Journal of Biometeorology*, vol 44, pp198–203

Pongracz, R., Bartholy, J. and Dezso, Z. (2006) 'Remotely sensed thermal information applied to urban climate analysis', *Advances in Space Research*, vol 13, no 12, pp2191–2196

Robitu, M., Musy, M., Inard, C. and Groleau, D. (2006) 'Modeling the influence of vegetation and water pond on urban microclimate', *Solar Energy*, vol 80, pp435–447

Sánchez de la Flor, F. and Alvarez Dominguez, S. (2004) 'Modelling microclimate in urban environments and assessing its influence on the performance of surrounding buildings', *Energy and Buildings*, vol 36, pp403–413

Santamouris, G., Mihalakakou, N., Papanikolaou, N. and Assimakopoulos, D. N. (1999) 'A neural network approach for modelling the heat island phenomenon in urban areas during the summer period', *Geophysics Research Letters*, vol 26, no 3, pp337–340

Santamouris, M. (2001) *Energy and Climate in the Urban Built Environment*, James and James, London

Santamouris, M., Papanikolaou, N., Livada, I., Koronakis, I., Georgakis, C. and Assimakopoulos, D. N. (2001) 'On the impact of urban climate to the energy consumption of buildings', *Solar Energy*, vol 70, no 3, pp201–216

Santamouris, M., Adnot, J., Alvarez, S., Klitsikas, N., Orphelin, M., Lopes, C. and Sanchez, F. (2004) *Cooling the Cities*, Eyrolles, Paris

Santamouris, M., Paraponiaris, K. and Mihalakakou, P. (2006) 'Estimating the ecological footprint of the heat island effect over Athens, Greece', *Climatic Change* (in press)

Sarrat, J., Lemonsu, A., Masson, V. and Guedali, D. (2006) 'Impact of urban heat island on regional atmospheric pollution', *Atmospheric Environment* (in press)

Shahoedanova, T., Burt, P. and Davies, T. D. (1997) 'Some aspects of the three dimensional heat island in Moscow', *International Journal of Climatology*, vol 17, pp1451–1465

Stathopoulou, E., Mihalakakou, P. and Santamouris, M. (2006) 'On the impact of temperature on tropospheric ozone concentration levels in urban environments', submitted for publication

Synnefa, A. and Santamouris, M. (2006) 'Development and performance of cool colored coatings', *Proceedings of Conference EUROSUN*, Edinburgh

Synnefa, A., Santamouris, M. and Livada, I. (2006a) 'A study of the thermal performance of reflective coatings for the urban environment', *Solar Energy*, vol 80, August, pp968–981

Synnefa, A., Santamouris, M. and Apostolakis, K. (2006b) 'On the development, optical properties and thermal performance of cool colored coatings for the urban environment', *Solar Energy* (in press)

Synnefa, A., Dandou, M., Santamouris, M. and Tombrou, M. (2006c) 'Cool colored coatings for passive cooling of cities', *Proceedings of EPEQUB Conference*, Milos, Greece

Szegedi, S. (2006) 'Heat islands in small and medium sized towns in Hungary', *Proceedings of Sixth International Conference on Urban Climate*, Gothenburg, Sweden

Szegedi, S. and Kircsi, Y. (2003) The development of urban heat island under various weather conditions in Debrecen, Hungary, available at www.geo.uni.lodz.pl/~icuc5/text/O_3_2.pdf

Szymanowski, M. (2005) 'Interactions between thermal advection in frontal zones and the urban heat island of Wroclaw, Poland', *Theoretical Applied Climatology*, vol 82, pp207–224

Taesler, R., Andersson, C., Nord, M. and Gollvik, L. (2006) 'Analyses of impacts of weather and climate on building energy performance with special regard to urban climate characteristics', *Proceedings of Sixth International Conference on Urban Climate*, Gothenburg, Sweden

Tayanc, M. and Toros, H. (1997) 'Urbanisation effects of regional climate change in the case of four large cities in Turkey', *Climatic Change*, vol 35, pp501–524

Theodosiou, T. (2003) 'Summer period analysis of the performance of a planted roof as a passive cooling technique', *Energy and Buildings*, vol 35, pp909–917

Tumanov, S., Stan-Sion, A., Lupu, A., Soci, C. and Oprea, C. (1999) 'Influences of the city of Bucharest on weather and climate parameters', *Atmospheric Environment*, vol 33, pp4173–4183

Unger, J. (1996) 'Heat island intensity with different meteorological conditions in a medium-sized town: Szeged, Hungary', *Theoretical Applied Climatology*, vol 54, pp147–151

Von Stulpnagel, A. (1987) *Klimatische Veränderungen in Ballungsgebieten unter besonderer Berücksichtigung der Ausgleichswirkung von Grünflächen, dargestellt am Beispiel von Berlin (West)*, PhD thesis, TU Berlin, Berlin

Von Stulpnagel, A., Horbert, M. and Sukkop, H. (1990) 'The importance of vegetation for the urban climate', in Sukopp, H. (ed) *Urban Ecology*, SPB Academic Publishing, The Hague, pp175–193

Watkins, J., Palmer, J., Kolokotroni, M. and Littlefair, P. (2002a) 'The London heat island: Results from summer time monitoring', *Building Services Engineering Research and Technology*, vol 23, no 2, pp97–106

Watkins, J., Palmer, J., Kolokotroni, M. and Littlefair, P. (2002b) 'The balance of the annual heating and cooling demand within the London urban heat island', *Building Services Engineering Research and Technology*, vol 23, no 4, pp207–213

Yaggie, C., Zurita, E. and Martinez, A. (1991) 'Statistical analysis of the Madrid urban heat island', *Atmospheric Environment*, vol 25B, no 3, pp325–332

Zanella, G. (1976) 'Il clima urbano di Parma', *Rivista Meteorologia Aeronautica*, vol 36, pp125–146

Ziomas, I. C. (1998) 'The Mediterranean campaign of photochemical tracers – transport and chemical evolution (MED-CAPHOT-TRACE): An outline', *Atmospheric Environment*, vol 32, pp2045–205

7

Post-occupancy Evaluation and Thermal Comfort: State of the Art and New Approaches

Elke Gossauer and Andreas Wagner

Abstract

Building occupants are a valuable source of information on building performance as well as indoor environmental quality and their effects on comfort and productivity. A large number of different studies have been carried out over recent decades focusing on various aspects of the broad field of comfort, well-being and health at workplaces. Two main methodical approaches can be identified analysing the history of comfort research: laboratory tests in climate chambers, and field tests in running buildings. Advantages and disadvantages of both approaches are pointed out in this paper and some of the most well-known studies from Europe and the US are described with respect to their specific objectives and methodologies. No difference is made between pure comfort field studies and more general post-occupancy evaluations because the borderline is vague in most cases and the purpose of this paper is to collect as much useful information as possible on comfort research methods. A current study on overall comfort at workplaces introduces a new multivariate approach, including building-related and further contextual parameters.

■ *Keywords* – comfort research methods; post-occupancy evaluation; workplace; comfort parameters; thermal comfort; indoor climate; general satisfaction

INTRODUCTION

The question of what constitutes a 'comfortable' thermal environment has preoccupied people ever since they have had to cope with indoor climate conditions. Important aspects have been, for example, building-related illnesses (Kröling, 1985), the development or planning of air-conditioning systems, normative requirements and design guidelines. Energy efficiency in buildings, which is addressed in the European Energy Performance of Buildings Directive of 2001, adds significant weight to comfort and workspace quality issues. On the one hand, energy savings with conventional cooling or air-conditioning systems must not lead to less comfortable indoor conditions. On the other hand, new energy efficient building concepts and technologies featuring natural ventilation and passive cooling require a revision of comfort standards that were developed for air-conditioned buildings only. Further, appropriate policies are needed to combine excellent energy building performance and high workspace quality in commercial buildings under different climates.

The question of comfort has always been approached by tests with people in order to include subjective votes and correlate them with measured climate parameters. While standards on thermal comfort were exclusively based on laboratory tests for a long time, the demand for field studies gradually arose. For example, building monitoring and surveys were meant to improve the dialogue between all partners of the design and building process and enhance their responsibility for the performance of the building (Bordass and Leaman, 2005). Further, the interdependence between indoor and outdoor climate was investigated in running buildings to test the applicability of the existing standards and predictive tools (ISO 7730, 2005). It became obvious that different results are obtained by testing people in their real working environment when this is not air-conditioned. Cross-correlations could be shown in the prediction of air quality, overall comfort and self-estimated productivity, and assumptions for comfort criteria (clothing value, metabolic rate) did not coincide with those given in the standards (Brager and de Dear, 1998; de Dear, 2004; Wagner et al, 2006). The development of the resulting adaptive comfort models is addressed in detail in the paper by Humphreys, Nicol and Raja (Chapter 3 of this volume).

Besides thermal comfort parameters a lot more information about indoor environment quality can be obtained by field tests. Almost every building has performance deficiencies to some extent, particularly during and directly after the commissioning phase. Technical equipment for heating, cooling, ventilating, air conditioning and lighting has to be adjusted and adapted to occupants' needs, which could be done by post-occupancy evaluation alternatively, or in parallel, to monitoring relevant data points. Post-occupancy evaluation is therefore a good opportunity to create a feedback loop for architects, planners and building industry professionals in order to learn how different building design features and technologies affect occupant comfort, satisfaction and productivity.

Despite an increasing interest in building performance assessment and post-occupancy evaluation, the results of such assessments are not routinely available. Further research on methodical approaches is necessary to finally implement post-occupancy evaluation as a standardized procedure into facility management. This paper gives an overview and evaluation of studies that have been carried out over recent decades in terms of objectives, methods applied and results. Although different definitions for post-occupancy evaluation and field studies of thermal comfort can be found in the literature (for example Nicol and Roaf, 2005), it is often difficult to draw a clear line between them in terms of purpose and methodology. Therefore in this paper post-occupancy evaluation is understood in a broad sense as an evaluation by people in buildings on the basis of questionnaires of various extent and generally in connection with physical measurements of different levels of detail.

BRIEF HISTORY OF OCCUPANT EVALUATION
AND COMFORT DEFINITIONS

The spectrum of scientific disciplines working on thermal comfort has been diverse, ranging from environmental psychologists dealing with perceived comfort and productivity in buildings (Cooper, 2001; de Dear, 2004), to engineers who always dominated research in this field. The first project on indoor environmental psychology and ergonomics had the aim of raising workforce productivity by manipulating the physical environment of workers. In the 1920s experiments with lighting levels were made in Hawthorne, US, by the National Research Council and it was found that any changes had an impact on productivity (Thommen, 2002). One of the results of these early experiments was that placebo changes also had a psychological effect on the workers. Therefore the assumption was made, that any positive intervention of the management was more important for productivity improvement than the actual environmental conditions. Probably due to this reason the issue was passed from psychologists to engineers, architects and end-users (Canter et al, 1975).

Single studies have been undertaken over the years, for example the Bedford study in 1936 (Bedford, 1936), but all in limited contexts. In 1963, the Royal Institute of British Architects (RIBA) published its Plan of Work for Design Team Operation (Bordass, 2005), which included occupant surveys and their feedback, because already at that time the need for evaluation in real existing buildings was recognized. This attempt was unfortunately withdrawn in 1972, probably because nobody was found to take over the costs of this service (Cooper, 2001).

The issue of building monitoring arose again in the 1970s, but then mainly due to problems of building-related illnesses (Kröling, 1985). At the same time the use of air conditioning (Brager and de Dear, 2003) and new materials was growing, which brought up the necessity of 'quantifying' comfort. It was Fanger in the 1970s (Fanger, 1970) who started systematic research on thermal comfort and the effect of clothing and activity upon the human heat exchange for that purpose. Since the 19th century, the understanding of comfort had only been related to the factors of light, heat and ventilation (Brager and de Dear, 1998). The newly developed heat balance models of Fanger (1970) and Gagge et al (1986) viewed people as passive recipients of thermal stimuli. This implied a quite simple cause-and-effect approach, where the physical boundary conditions of the indoor environment influenced the human physiology and therefore the possible conditions of subjective comfort or discomfort.

The succeeding standards, based on the heat balance model of the human body and derived from extensive experiments in climate chambers, were appropriate for buildings with air conditioning, though they suggested that they were applicable for all building types, population and climates (Fanger, 1970). As a result, clear limits for the indoor climate were given that had to be met by the heating, ventilation and air-conditioning (HVAC) systems.

Field studies at that time already recognized the interaction between people and their surroundings, for example by changing their clothes, their work speed or opening/closing their windows (Nicol and Humphreys, 1973) (Figure 7.1). This already implied the theory that people can adapt to their thermal environment.

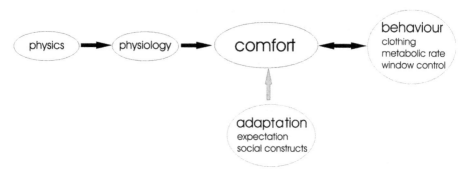

Source: Elke Gossauer
FIGURE 7.1 Cause-and-effect approach for field studies of thermal comfort during the 1970s

The 1980s brought further changes in the building industry and other attempts to establish post-occupancy evaluation were made in several countries. There was great competition within the building industry to improve the speed and quality of the building process and to reduce the costs of building materials. The HVAC industry grew rapidly and also new requirements of information technology had to be adopted. Since then, buildings have been used more intensively and are strongly influenced by social, technical and marketing changes.

One of the effects of this was the discussion and extension of the definition of thermal comfort. For example Cooper (1982) claimed that comfort standards were 'social constructs, which reflect the beliefs, values, expectations and aspirations of those who construct them'. In the 1980s it was argued that thermal comfort is a multivariate phenomenon that is influenced by behaviour (clothing, activity and ventilation rate) and expectations, as well as by environment and memory (Brager and de Dear, 2003) (Figure 7.2).

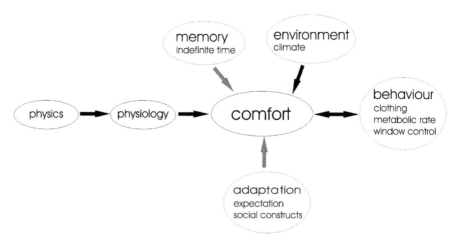

Source: Elke Gossauer
FIGURE 7.2 Cause-and-effect approach for field studies of thermal comfort in the 1980s

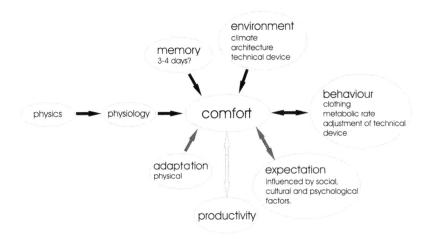

Source: Elke Gossauer

FIGURE 7.3 Cause-and-effect approach for field studies of thermal comfort in the 1990s

In the ASHRAE standard of 1992 the definition of comfort was established. Comfort was defined as 'the condition of mind that expresses satisfaction with the thermal environment'. With the growing complexity of the indoor environment, it became almost impossible to 'measure' comfort directly (Brager and de Dear, 2003). Besides measuring the physical variables that influence the body's heat balance, questions about thermal sensation and preference were introduced, both in the climate chamber and in the field. Conclusions about satisfaction or dissatisfaction of people were derived from that. With expanding research, further topics like architecture and technical devices, as well as social, psychological and cultural factors were added and confirmed (Nicol and Kessler, 1998) (Figure 7.3).

In the 1990s, interest rose in finding correlations between different indoor parameters and the perception of health, comfort and productivity at work. More and more field surveys, mostly independently targeted, were realized in different countries to analyse different comfort parameters and how people react to changes of their indoor environment (Leaman and Bordass, 2001). A very good review of projects until 1997 is given in the literature review of Brager and de Dear (1998).

Today a stronger emphasis is given to psychological parameters and their impact on satisfaction and productivity, but also to possibilities of energy saving in buildings while maintaining a high comfort standard. Besides the physical parameters, the following criteria have been identified to have a strong impact on occupant satisfaction with temperature (de Dear et al, 1997):

● adjustments of clothing, blinds, windows, heating and so on;
● physiological acclimatization and psychological expectation;
● thermal expectation formed by the past thermal history, weather forecast and so on; and
● time of exposure and non-thermal factors such as available or performed control.

Although many studies are not mentioned in this review, this brief history of evaluation methods and investigated topics clearly illustrates the (growing) complexity of the interrelations between building-related factors and their influences on occupants' (thermal) comfort. Yet even now, many correlations have not been found or confirmed. Multidisciplinary work is therefore necessary and multiple methods have to be connected. Both possible approaches for the evaluation of (thermal) comfort in buildings – research in the laboratory and in the field – are discussed in more detail below.

WHAT ARE THE DIFFERENCES OF OCCUPANT EVALUATION IN THE LABORATORY AND IN THE FIELD?

Thermal comfort studies have been carried out intensively since the 1970s under laboratory conditions. Field studies have been conducted too, but with different emphases, as described above. In climate chambers, physiological and psychological reactions of human subjects to climate parameters, such as air temperature, radiant temperature, humidity and air velocity, can be reproducibly investigated under controlled conditions. Together with a clothing value and the metabolic rate, Fanger validated his predicted mean vote–percentage of people dissatisfied (PMV–PPD) model (Fanger, 1970) with these experiments for defined indoor conditions. This model forms the basis of the ISO 7730. The model implies a steady-state human heat balance, which is independent of external climate parameters. It was not possible for the subjects to change the climate conditions they were experiencing or their clothing (Brager et al, 2004).

Laboratory-based studies are likely to ignore contextual influences that can weaken the meaning of responses to a given set of thermal conditions. From chamber studies, it has not yet been possible to assess how dissatisfactions from multiple sources are combined because they are masked. Moreover, climate chambers generally do not adequately reflect a workspace environment that occupants experience in real buildings.

Because of increasing doubts about the external validity and relevance of climate chamber results for building occupants over recent decades, ASHRAE commissioned a series of thermal comfort studies in the 1990s. The aim was to validate the findings of experiments in climate chambers and the resulting standards for HVAC systems in the field in a variety of climatic contexts around the world (de Dear, 2004).

Occupants' responses to conditions in real buildings may be influenced by a range of complex factors that are not accounted for in the heat balance models. Field studies consider the whole indoor environment of the surveyed people, depending on the survey methods. Accordingly it is possible to analyse how different building types and technologies react to the outdoor climate and how this is perceived by the building occupants.

A major problem of field studies is that environmental conditions in most buildings are transient and difficult to measure accurately, together with the occupants' votes. Therefore 'errors of the input data give rise to errors in the relationship in statistical analysis' (Humphreys and Nicol, 2000). Care has to be taken in generalizing the results of one survey from statistical analysis, even if the circumstances in other buildings are similar.

Another challenge of a multicriteria approach in real buildings is that the correlations between multiple parameters still have to be found and validated. This is similar to the analysis of the sick building syndrome and productivity research. However, more recent and systematically performed field studies and experiments give the basis for quantifying the specific causal mechanisms concerning thermal comfort and adaptation of occupants. One example is the adaptive comfort model that gives a relation between the predictors for thermal comfort known from the laboratory and the outdoor temperature (Brager and de Dear, 1997; de Dear, 2004).

With ongoing evaluation in the field, it was becoming increasingly evident that the explanation of thermal comfort is not as distinctive as thought on the basis of research in the laboratory. Recent studies report that other comfort parameters, such as indoor air quality or visual and acoustic conditions, correlate with thermal and overall comfort and well-being (Roulet et al, 2006a). The ability to control indoor environmental parameters influences satisfaction with temperature. Differences in the perception of noise and sounds were found in relation to social and cultural statuses and different relations to the source. For further analysis of thermal comfort, discussion must be widened beyond the experimental environment (climate chamber or running building) to include other aspects of experimental design.

CLASSIFICATION OF SURVEY PROCEDURES AND RATING SCALES
Brager and de Dear (1998) describe three classes of thermal comfort studies:

- Class I: All sensors and procedures are in 100 per cent compliance with the specifications in ASHRAE 55 (ASHRAE, 2004) and ISO 7730. Measurements at three heights above floor level with laboratory grade instrumentation are required. This procedure allows a careful examination of the effects of non-uniformities in the environment as well as a comparison between buildings.
- Class II: All physical environmental variables necessary for calculating PMV and PPD indices are measured and collected at the same time and place when and where the thermal questionnaires are administered, most likely at one height. This allows an assessment of the impact of behavioural adjustment and control on subjective responses.
- Class III: This is based on simple measurements of indoor temperature and possibly relative humidity at one height above the floor. The physical measurements can possibly be asynchronous with subjective measurements (questionnaire with rating scales).

Most of the field studies conducted over recent decades have been Class III studies. The quality does not necessarily allow explanatory analyses by extensive statistics. The analysis is limited by its ability to determine causal mechanisms, but gives a good and generally quick overview over the situation and some correlations within and between the buildings. Particularly in naturally ventilated buildings, measurements and the filling of the questionnaires should take place at the same time because there are rooms that might have temporarily or spatially non-uniform thermal conditions.

Rating scales for thermal sensation have been in use for over 50 years, others even longer. Details for performing measurements in the field are given in the existing standards (ASHRAE and ISO 7730). Measurement parameters and survey questions are mostly adopted from research in the laboratory. With the ASHRAE scale (in the US) a person is thermally comfortable at the neutral point of no thermal sensation, and so thermal comfort is here defined as the absence of thermal discomfort. The other scale that is commonly still in use is the Bedford scale (in the UK). The Bedford scale confounds warmth and comfort, which makes it less applicable when compared to the ASHRAE. Other scales are also used.

MAIN STUDIES AND THEIR OBJECTIVES

Over the last two decades, field studies have been performed with different objectives and therefore by means of partly varying methodologies, mainly by researchers from universities. However, companies like BREEAM (BRE's Environmental Assessment Method) in Britain and LEED (Leadership in Energy and Environmental Design) in the US were established to commercially investigate comfort as part of the sustainability of buildings (benchmarking of 'green buildings').

In 1981 Building Use Studies (BUS) was founded in London as an independent consultancy (Bordass and Leaman, 2005). Since then BUS has examined how buildings perform, particularly from the viewpoints of building occupants. The company has carried out post-occupancy evaluations in the UK and around the world, investigating occupants' health, perceived comfort (including personal control) and self-estimated productivity. One of the prevalent objectives has been a benchmarking of the analysed buildings. One of the BUS projects was an evaluation of 40 office buildings, comparing the self-estimated productivity of occupants between naturally ventilated and air-conditioned buildings (Leamann and Bordass, 1999) in connection with the depth of buildings and office zones respectively. Feedback to the users is seen as one of the most important issues of BUS building monitoring. Many outputs from BUS (though not commissioned consultancy reports) are available via the website, www.usablebuildings.co.uk. BUS survey methods are available under licence and are in use worldwide.

The interest in differences between naturally ventilated and air-conditioned buildings is almost as old as air conditioning itself. Research topics have been, for example, building-related illness, perception of thermal comfort and indoor air quality and later on, self-estimated productivity. At the end of the 1990s a series of thermal comfort studies was performed (Nicol and Kessler, 1998; Rowe, 1998) that mainly investigated the influences of the ventilation type of the building, the proximity of occupants to windows and their access to natural ventilation facilities on the rating on thermal and overall comfort. Research by de Dear and Brager (De Dear et al, 1997) revealed that occupants of naturally ventilated buildings are comfortable in a wider range of temperatures than occupants of buildings with centrally controlled HVAC systems (Mendell, 1993). The comparison of different field studies in four countries led to the development of the adaptive comfort model. However, the exact influence of personal control in explaining these differences could only be hypothesized because of the limits of the existing field study data that formed the basis of that research.

Therefore further projects were initialized to quantitatively investigate how personal control of operable windows in office settings influences local thermal conditions and

TABLE 7.1 Post-occupancy studies mainly related to research in thermal comfort
and its predictive parameters

PROJECT NAME	YEAR	BUILDING TYPES	OBJECTIVES
SCATS	1997–2000	Naturally ventilated (NV), air-conditioned (AC)	Correlation between comfort temperature and indoor/outdoor temperatures, behavioural analyses
PROBE	1995–2002	NV, AC, mixed	Energy and environmental performance, thermal comfort, occupant satisfaction, feedback
RP-1161	Unknown	NV, AC	Influence of personal control on thermal comfort, self-controlled acquisition of physical and subjective data

occupant comfort (PROBE, 1999; Nicol and McCartney, 2000; Bischof et al, 2003). Their aim was to disentangle the precise effect of personal control from other potential explanations for people's acceptance of more variable thermal conditions. The data showed that occupants with different levels of personal control had significantly diverse thermal responses, even when they experienced the same thermal environments and clothing and activity levels.

Another principal concern of those studies was to relate thermal comfort and the comfort temperature to the indoor and outdoor climate and to compare them with the PMV index (Table 7.1). The data of all three studies include a wide range of subjective and environmental measurements and the database (with both environmental and subjective data) has mainly been used to define an adaptive algorithm but also for the analysis of technical and environmental performance of the buildings, occupant satisfaction and productivity (Leamann and Bordass, 1999). The projects also showed a more systematic approach to the surveys.

Two projects investigating the issue of sick building syndrome were realized in Germany and Switzerland (Table 7.2). Again, the influence of technical features (natural ventilation or air conditioning) was one of the main interests.

In the ProKlima study (Bischof et al, 2003), 14 German office buildings were evaluated with regard to sick building syndrome in air-conditioned and free-running

TABLE 7.2 Post-occupancy studies mainly related to research in the field of sick building syndrome

PROJECT NAME	YEAR	BUILDING TYPES	OBJECTIVES
ProKlima	1995–2003	NV, AC	Contribution of the indoor climate, energy concept and psychological factors to the illness symptoms and thermal comfort
HOPE	2002–2005	NV, AC	Benchmarking of 'healthy' and energy efficient buildings, input into CEN standards

buildings. The study, which was realized by six interdisciplinary German research groups, was the first one with a distinct experimental design and systematics in the field of sick building syndrome. By means of a questionnaire, physical measurements and medical investigations, the impacts of the indoor climate, as well as psychological factors, on illness symptoms were evaluated. Hellwig (2005) further analysed the existing data to reveal correlations between ventilation strategies and thermal comfort.

The objective of the Health Optimisation Protocol for Energy-efficient Buildings (HOPE) project was to define a set of qualitative (prescriptive) and quantitative (measurable) performance criteria for healthy and energy efficient buildings in different European climates. Results should directly be incorporated into European Committee for Standardization (CEN) standards and in this context be used as guidelines for improving building performance. The study mainly produced a coarse benchmarking of buildings (green, yellow, red and black labels) by using basic statistics, for example, frequency distributions and assigning a British Standards Institution (BSI) Index in connection with air-conditioned and naturally ventilated buildings by the analysis of variance. Data are still being analysed (Bordass and Leaman, 2004; Roulet et al, 2006 b;HOPE, 2006).

A recently growing interest of post-occupancy evaluation is the quality benchmarking of buildings as a feedback instrument to architects, engineers and industry, and also to the occupants (Table 7.3). For this purpose the Center for the Built Environment (CBE) in Berkeley has used indoor environment quality (IEQ) surveys since 1996 to systematically collect and archive occupant responses in buildings. Their work includes building diagnoses, evaluation of new building technologies, identification of new trends in building performance, and benchmarking the quality of individual buildings against the population of similar buildings. By 2005 surveys in more than 70 buildings had been finished (including office buildings, laboratories, banks and courthouses) and a rapidly growing database of standardized survey data was established that is used for benchmarking (Zagreus et al, 2004) and advanced data analysis. The IEQ surveys are often part of a post-occupancy evaluation process in which the design and operation of a building is assessed. Different applications are available, for example, for pre- or post-analysis of occupants moving into a building, correlations of occupants' ratings with physical measurements, and evaluation of clients' design objectives. Feedback to building managers, occupants or industry partners is given by means of a self-controlled reporting tool.

TABLE 7.3 Post-occupancy studies for benchmarking and performance feedback

ORGANIZATION	YEAR	BUILDING TYPES	OBJECTIVES
CBE, Berkeley	Since 1996	NV, AC, mixed	Diagnosis of problems, evaluation of new building technologies, quality benchmarking
CCC	Since 2001		Feedback on the performance of industry products for buildings
BREEAM, LEED	Unknown year of start until today		Benchmarking of 'green' buildings

In 2001 the Confederation of Construction Clients (CCC) was formed in the UK for better cooperation between clients and their suppliers. CCC started a research project developing a feedback system that should help to obtain better information about the performance of completed buildings. The activities of BREEAM and LEED, detailed above, also form a contribution.

APPROACHES TO POST-OCCUPANCY EVALUATION IN DIFFERENT STUDIES

In all studies a crucial point was to establish from the outset is good contacts with all persons in authority to ensure cooperation between occupants and the project team. The procedures followed during a majority of building evaluations (Figure 7.4) are outlined below.

An inspection of each building is necessary to provide basic data of the buildings and their environment. This is commonly followed by interviews with the building management, from which information on building energy performance (and other issues) is collected. As a third step a questionnaire is distributed to the occupants – either on paper or via internet or intranet (either web or email) – that gathers information on how the internal environment is perceived and rated. Measurements of physical parameters are performed according to the different methods outlined above.

De Dear et al (1997) collected and analysed a database (ASHRAE RP-884) consisting of data derived from thermal comfort field experiments in 160 buildings (mainly commercial offices) from four continents, which were either naturally ventilated (45) or had centrally controlled HVAC systems (111). The number of investigated buildings per study ranged from 1 to 16. The occupant samples per building lay between 6 and 380 persons, and approximately 22,000 comfort questionnaires were completed in total.

Some studies were divided into summer and winter surveys. One requirement of the surveys was that the occupants' microclimate at their workplace be simultaneously monitored. Therefore a detailed set of indoor climate and thermal comfort data is available for each building.

In most of the surveys the findings were rated on the ASHRAE seven-point scale. Sometimes the Bedford Scale was also used. Only in some of the studies was additional information requested on either thermal acceptability or preference, and seldom for both. Where preferences were addressed, three-point or five-point scales where used to rate them. Clothing values were either estimated by ISO 7730 or ASHRAE 55, sometimes taking into account the occupant's chair.

The ASHRAE-RP-1161 project was conducted in a totally naturally ventilated building in Britain. The office layout was mainly open-plan, with varying levels of direct or indirect individual control, based on proximity of occupants to the operable windows. Continuous measurements of the microclimate at each workstation were taken and the occupants had to fill in a web-based questionnaire that was developed at the CBE. Both a general background survey and a short repetitive survey were conducted. The questionnaire for general background was administered to all 230 occupants and included basic demographics, personal workplace characteristics and aspects such as individual environmental control opportunities, window operation and satisfaction with various environmental attributes. The return rate was around 40 per cent, that is, the norm for web-based surveys (Zagreus et al, 2004).

In a second step, 38 persons (24 females, 14 males), all of them volunteers, had to fill in a short repetitive questionnaire evenly distributed over each day after having been at their workstation for at least 30 minutes. The survey was linked with indoor climate

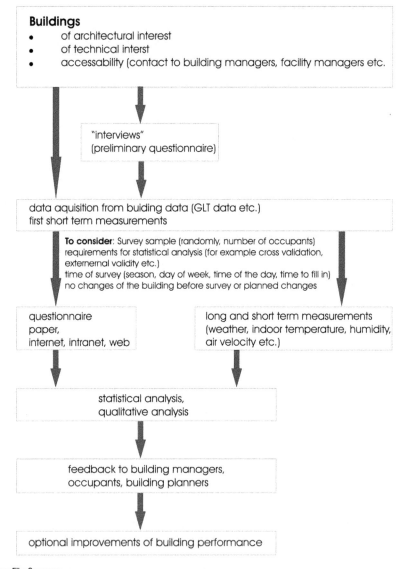

Source: Elke Gossauer

FIGURE 7.4 General procedure of building surveys

TABLE 7.4 Field experiment methodology of ASHRAE RP-1161

	BACKGROUND STUDY	DETAILED STUDY
Physical measurements	Continuous measurements of indoor climate conditions: temperature and humidity in different zones within the building	Continuous measurements of workplace climate conditions: dry-bulb temperature, globe temperature, air velocity, plus nearby meteorological station
Web-based survey	Once, before detailed survey, seasonal impressions, 10–15 minutes long	Several times/day, current impressions, 1–2 minutes long
Subject pool	Entire building occupant population	38 voluntary subjects
Duration and frequency	Two-week period during winter and summer; subjects took the survey only once, at their convenience, during that period	Two-week period in each of the two seasons immediately following the background study; subjects took the survey repetitively during that period, on average two to three times per day

measurements that were taken with very similar instrumentation to that used in laboratories. Measurements and short repetitive surveys took place over two weeks in winter (February and March 2003) and two weeks in summer (September and October 2002). The occupants had to fill in a list of clothing options and five different descriptions of office tasks to describe their activity during the previous 30 minutes for the calculation of the clothing values (including the chair) and the metabolic rate. In addition to traditionally used questions, newly developed questions about thermal variability, air movement, window and blind patterns and other environmental adjustments, as well as energy level and mental alertness were added. Over 1000 survey responses in each of the two main seasons were collected and the subjective and physical measurements were cross-linked. In Table 7.4 the methodology of the surveys and measurements of the ASHRAE RP-1161 project are shown.

The SCATs (Smart Controls and Thermal Comfort) Project included the evaluation of 26 office buildings situated in England, Sweden, Portugal, Greece and France (Nicol and McCartney, 2000) (Table 7.5). Each partner of the research team chose five buildings for the surveys. The aim was to have some 200 subjects per country over 12 months of surveys, but this was not always achieved (850 people finally took part in the surveys). The number of subjects per building taking part in the full monthly surveys (Level 1) varied from 4 to 111. A background survey was conducted in each of the buildings in which all occupants were asked to fill in a paper questionnaire. The intention of this questionnaire was to explore the attitude of the subjects to the building they were working in and to give information about their job, the working environment, individual control possibilities of the

TABLE 7.5 Field experiment methodology of the SCATs project

	BACKGROUND STUDY	DETAILED STUDY
Physical measurements		Transverse study: CO_2, globe temperature, air temperature, relative humidity, illuminance, air velocity and noise level. Longitudinal study (local): temperature (and humidity)
Survey	Background survey once per building	Transverse questionnaire; longitudinal questionnaire
Subject pool	4 to 100 people per building occupant population	38 (33) people
Duration and frequency	Within 12 months	Once each month and four times per day over 2–12 months

indoor climate, health and personal details. Completed questionnaires were returned by 40 per cent (352) of the subjects. A transverse questionnaire was also administered monthly to a subset of persons for up to a year. This was accompanied by detailed physical measurements that included CO_2 concentration, globe temperature, air temperature, relative humidity, illuminance, air velocity and noise level. A longitudinal questionnaire was then filled in up to four times a day by 33 persons in all buildings. The temperature (and in some cases the relative humidity) close to the subject was recorded by a miniature data logger.

The comfort vote was rated within a seven-point scale. Additionally a five-point preference scale was used. Topics were temperature, air movement, humidity, lighting, noise, air quality, overall comfort and perceived productivity. In the transverse questionnaire a six-point scale for the rating of overall comfort was used. Details of clothing and metabolic rate were also collected using shortened versions of the clothing and activity descriptions from the longitudinal study. The clo value was calculated from standard tables, including a value for the chair. For evaluating the metabolic rate, activities over the previous hour were recorded. The weighted mean metabolic rate was calculated for all the reported activities. The actuation of controls of doors, heating, air conditioning, windows, blinds, lights, fans and so on at the time of the survey was also taken into account in the analysis of occupants' behaviour. In the French survey, an electronic form of the questionnaire was used frequently. Some statistical analysis of environmental data (for example, standard deviations) had been incorporated into the self-controlled analysis, which could be further developed in the future.

During the PROBE (Post-occupancy Review of Buildings and their Engineering) survey, about 20 buildings with interesting technical features, typically completed between two and five years previously, were investigated (PROBE, 1999; Bordass and Leaman, 2004) (Table 7.6). A pre-visit questionnaire of about five pages was handed out to the facilities managers to get information on the building, its operation and its utility consumption, but this was seldom filled in. The survey method that was used in the PROBE study was first developed and used in the 1980s by BUS in a comprehensive study on sick building

TABLE 7.6 Field experiment methodology of the PROBE project

	BACKGROUND STUDY	DETAILED STUDY
Physical measurements	Spot measurements of light, temperature and relative humidity	Light, temperature, relative humidity, electricity and heat supply, pressure tests
Paper survey	Pre-visit questionnaire for facilities managers Background survey: once per building	Detailed paper questionnaire
Subject pool	One person	Approximately 100 persons/building
Duration and frequency	Some weeks	Once during one day

syndrome. An energy assessment spreadsheet was developed to prepare benchmarks of buildings. This was further developed into the CIBSE TM22 energy survey method in 1998 together with associated software. For PROBE the questionnaire was reduced in length to be speedy, easy and attractive to use and analyse. The occupant questionnaire consisted of about 40–50 variables within 12 topics that covered personal statistics, overall building aspects, individual control, speed and effectiveness of management response after complaints have been made, temperature, air movement, air quality, lighting, noise, overall comfort, health, productivity at work and some site specific questions. Most of the questions had to be answered within a seven-point 'Gregory' scale and space for personal comments on specific and general issues was provided. No questions about occupations were included because of objections by most building managers. Only one question on health and self-estimated productivity was included. Recently questions about cleaning and furniture were added. A shorter secondary questionnaire was given to special user groups, most commonly students in educational buildings.

The questionnaire was typically administered to a sample of 100–125 permanent staff or to everybody in buildings with less than 100 occupants. The sample size was important for statistical validity and to permit analysis of sub-samples within a building. The return rate was typically above 90 per cent due to the questionnaire being personally distributed by BUS members, thus giving occupants an opportunity for personal discussion. They also received all relevant information about the purpose and date of the survey in advance. The questionnaires were collected on the same day.

The analysis included two summary indexes: one based on comfort, compiled from scores for perceived summer and winter temperature and air quality, lighting, noise and overall comfort; the other based on satisfaction, using scores for design, needs, productivity and health. Recently a third index was formed at the request of clients to combine the first and second.

Spot measurements of light, temperature, relative humidity, electricity and heat supply were taken at the first site visit. Sometimes data were available from site management or independent monitoring. Pressure tests were conducted due to air tightness problems in many of the buildings. Each building was then benchmarked against a broader data set.

TABLE 7.7 Field experiment methodology of the ProKlima project

	BACKGROUND STUDY	DETAILED STUDY
Physical measurements	Indoor air quality, noises, thermal comfort parameters (every 15 minutes), light, ergonomics of workplaces	Indoor air quality, noises, thermal comfort parameters, light, medical investigations, test of concentration
Survey	Background survey	Detailed questionnaire
Subject pool	At least 200 persons	120 persons per building
Duration and frequency	Investigations took place every day during one week, survey was conducted once	Several weeks with building investigations, physical measurements, medical investigations and survey

The benchmarks are available and largely published or available under licence (Leaman and Bordass, 2001).

For the ProKlima project, all participating buildings had to have more than 200 occupants undertaking 'typical' office work (Table 7.7). Also no buildings with changes in construction during the previous two years were allowed to participate. Eight of the buildings had different types of air conditioning and six were naturally ventilated. 1500 workplaces were analysed with the same number of occupants being questioned and medically investigated. Indoor air quality, noises, thermal comfort parameters and light were measured during the study. Temperatures and humidity were measured at one height (1.10m) over a period of 15 minutes in each building. Tests of concentration (Bordass and Leaman, 2004) were conducted at the computer. Psychological factors were also considered to have influence. After having asked all occupants in the building, a sample of 120 people per building was chosen (60 subjects with symptoms, 60 without symptoms) and measurements were taken at their individual workplaces. Paper questionnaires were distributed that included seven pages of questions about physical perceptions, health, well-being, indoor climate (including odours, light, noise, temperature), satisfaction with work and personal statistics. The questions about the indoor climate also included five-point scales about satisfaction and its importance for general well-being. The ability of individuals to influence indoor climate conditions had to be answered by 'yes' or 'no'. For evaluating sensory aspects as well as the indoor climate, a new procedure of questions was developed and validated for this study. The return rate was between 73 and 90 per cent per building.

During the HOPE project, 64 office buildings (and about 111 dwellings) in nine countries were investigated (HOPE, 2006) (Table 7.8). The buildings that were included in the study were not distributed representatively among the various European climates. The main criterion for each building to be included in the study was that access to basic information on design, building fabric, services and so on, was available. Further, reliable information about energy use for a minimum of 12 months was required to provide specific energy consumption data, including climate data from locations reasonably close to each building. About 75 per cent of the buildings showed a low energy standard based

TABLE 7.8 Field experiment methodology of the HOPE project

	BACKGROUND STUDY	DETAILED STUDY
Physical measurements	Energy use collected for several years	VOCs, asbestos, NO_x, CO_2 and radon
Survey	Building checklist	Questionnaire
Subject pool		On average 90 per building
Duration and frequency		Once

on regular records in buildings or energy bills, which were collected for several years and then averaged. No correction was made for the different climate regions.

At a minimum, 50 subjects per building were required for the survey. Only buildings without changes in technology or architecture for a minimum of one year prior to the start of the study were admitted. For the more detailed study, no major renovation planned before autumn 2004 was allowed either. The questionnaires focused on occupants' satisfaction in terms of comfort and their perceived health within the building. The questionnaire was prepared especially for the HOPE project. About 420 variables were gathered in each office building and approximately 6000 valid questionnaires were collected for analysis. Overall comfort was evaluated by equally taking into account the parameters of temperature and its variability, noise from the outside and from the building itself, natural and artificial lighting, and various criteria related to the air quality. These comfort parameters had to be rated for the summer and for the winter within one survey using a seven-point scale from satisfactory to unsatisfactory. Additional questions about possible control of the indoor environment were separately asked for temperature, ventilation, shading from the sun, lighting and noise. Additional measurements of volatile organic compounds (VOCs), asbestos, nitrogen oxides (NO_x), CO_2 and radon were also taken in the buildings.

To date, analysis of the data has focused on indoor environment quality and general building characteristics. Qualitative analysis involves sorting the buildings into 'best' and 'worst' categories and looking at the differences between them. All variables used in the study are mean values for each building. The building-averaged data were collected and compared, looking for correlations between building characteristics and perceived comfort and health. The aggregation method used for comfort variables assumes that all criteria – such as temperature, noise, light and air quality – have the same weight for the perception of overall comfort and health. The classification made for the study was not intended to be mandatory (Roulet et al, 2006a).

The CBE in Berkeley (http://cbe.berkeley.edu/research/publications.htm) have surveyed 142 commercial buildings (23,450 people) since the beginning of their project (Table 7.9). The core survey focuses on seven key areas of the indoor environment: office layout, office furnishing, thermal comfort, indoor air quality, lighting, acoustics, and building cleanliness and maintenance. The CBE has also developed additional question sets to gather information on specific aspects of the workplace environment. Examples of optional modules include way-finding, safety and security, operable windows, shading systems, floor diffusers and washrooms. Other modules can be optionally developed and included to investigate additional topics.

TABLE 7.9 Field experiment methodology of the CBE

	BACKGROUND STUDY	DETAILED STUDY
Physical measurements		Depending on the study, basically none
Web survey		Core questionnaire, additional modules
Subject pool		All building occupants, depending on client
Duration and frequency		5–12 minutes, open for 1–2 weeks

The survey is web-based and offered on the CBE server in different languages. The questionnaire can be answered within 5–12 minutes depending on the number of branching questions, whenever dissatisfaction is indicated. Ratings have to be given within a seven-point scale. The survey is open for approximately two weeks, sometimes combined with a reminder via email. Response rates range from 27 to 88 per cent, with an average of about 50 per cent. The questions and modules have been kept consistent over the years and data are collected in a SQL-Server database. Thus a broad analysis and comparison between buildings is possible. Self-generated reports are available immediately after a survey via a password-protected website. Generally no physical measurements are taken and the buildings are not inspected before or during the survey.

PRELIMINARY CONCLUSIONS FROM THE STUDIES

A lot of field studies have been performed over recent decades focusing on different issues and applying different methods of post-occupancy evaluation. In contrast to laboratory experiments, they gave a strong emphasis to a broad range of building- and workplace-related parameters – for example, ventilation strategy and technological features for energy efficiency – and their influence on thermal comfort. Despite the fact that field studies and laboratory experiments are based on different research methodologies, existing scales and measurement procedures of laboratory studies that are also part of current standards were used frequently (Brager and de Dear, 1997), especially for collecting votes on thermal comfort. Due to the growing awareness of the complexity of the issues, each project has developed additional questions with changing rating scales (both even and uneven, from three-point to seven-point). The spectrum of parameters to be investigated has also grown: office layout, workload, noises, light and more have been added to find out whether and how strongly they correlate with other environmental parameters.

By contrast, most of the studies concentrated on one, or at least on a reduced number of, specific topics within the wide range of possible analyses. This is due to the large amount of data that can be collected and the restricted manpower of the research teams. Occupant surveys are mostly an economical compromise between the objectives of a study, the needs of respondents (question-answering ability), data management, data analysis and statistical validity.

The statistical analysis in the occupant surveys considered P-values of less than 5 per cent (p0.05) to be significant. In addition to frequency distributions, mean values and correlations and in some studies also multiple regression analyses and analyses of variance were generally calculated. Though building and occupant samples were not the same and different rating scales, measurements and procedures for the analyses were used, similar results were frequently obtained in comparable studies. This is particularly true for aspects such as the importance of having individual control of the indoor climate at the workplace, the low relevance of satisfaction with light for overall satisfaction with the workplace, and the rating of the noise level in correlation with the source and surroundings (PROBE, 1999; Roulet et al, 2006b; Gossauer et al, 2006).

APPROACHING OVERALL COMFORT AND THERMAL COMFORT IN SUMMER BY NEW FIELD STUDIES

Since January 2004 a post-occupancy study has been carried out in 16 German office buildings (Gossauer et al, 2006). Within the study overall comfort at the workplace and its weighted influences by thermal, visual and acoustic comfort parameters, indoor air quality and the office layout are examined. The study also considers different influences on perceived health at work and the impact of work-related factors on satisfaction and perceived productivity. Approximately 70 questions have to be answered within a five-point Likert scale, once in winter (between the middle of January and the middle of March) and once in summer (in August or September). Thus, seasonal climate influences on the occupants' judgement, particularly on temperature and lighting votes, are taken into account.

For the study, a modified version of the CBE questionnaire was used that had been previously adopted in a survey in nine office buildings of the German Railway Company (DB Netz AG) (www.enerkenn.de). A copy of the modified questionnaire can be found in Voss et al (2005).

The survey was carried out anonymously with a sample size of 30–100 randomly chosen people per building (depending on the size of the building). An average return rate of more than 80 per cent was achieved by handing out paper questionnaires personally to the participants. Approximately 1400 questionnaires were included in the multivariate statistical analysis that began in June 2006.

Additionally, room temperatures and humidity were measured with portable data loggers on the day of each survey. In some of the buildings, more data (for example, continuously logged room temperatures, opening times of windows and indoor air quality) are available from different monitoring campaigns and will be used for further evaluation. Analysis of the occupants' responses is carried out with the statistical software program SPSS (Statistical Packages for the Social Sciences, Version 11.5 and 13.0). It includes the calculation of mean values, frequency distributions and correlation values, as well as a regression analysis for dependent factors. Furthermore the correlations between independent factors are considered, for example between the general satisfaction and individual satisfaction parameters. To identify significant differences in the ratings between summer and winter, an analysis of variance was carried out. The hypotheses were statistically tested with a two-tailed alpha level of 0.05; the different sample sizes

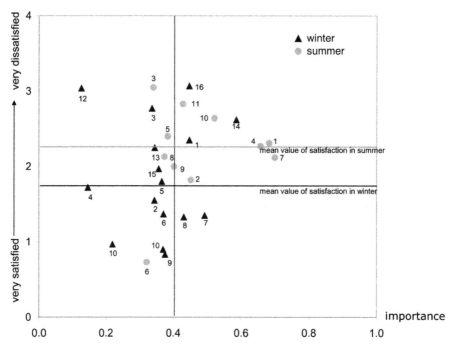

Note: The different numbers refer to the buildings in which the surveys took place. Field A (lower left sector): occupants are satisfied with the parameter but the weighting calculation shows that it is less important for the general satisfaction with the workplace. Field B (lower right sector): occupants are satisfied with the parameter and it is important for the general satisfaction with the workplace. Field C (upper left sector): occupants are dissatisfied with the parameter but it is of less importance for the general satisfaction with the workplace. Field D (upper right sector): occupants are dissatisfied with the parameter and it is very important for the general satisfaction with the workplace.
Source: Gossauer et al (2006)

FIGURE 7.5 Correlation between mean satisfaction with temperature and weighted importance of temperature for general satisfaction with the workplace (Spearman correlation)

and the occasional imparity of variance were considered as well. A cluster analysis was used to identify possible groupings of building characteristics.

For the evaluation, the extent to which individual satisfaction parameters influence general judgement of the workplace can be correlated with general satisfaction of the workplace. This leads to weighted values of the importance of each parameter in relation to the general satisfaction. This weighting procedure proved to be more reliable when compared to the occupants' judgements because occupants mostly tend to choose the categories 'important' or 'very important' if asked directly. Figure 7.5 illustrates the weighting of the satisfaction with temperature for the general satisfaction with the workplace. It can be seen that in summer dissatisfaction with room temperature is generally higher than in winter, and also the importance of this parameter for general satisfaction with the workplace is not necessarily exclusively dependent on the grade of dissatisfaction.

One aim of this study is to propose limits for the different comfort parameters concerning workplace satisfaction. Further objectives are to suggest standards for questions and scales and the statistical analysis for the evaluation of overall comfort. Thus a tool for a more straightforward building assessment can be provided. An innovation for this kind of study is the consistent seasonal evaluation of all buildings and the accurate evaluation of seasonal differences and coherencies between different parameters by means of multiple regression analysis and cluster analysis. A model for the validation of the item structure is currently being calculated and thus confirmation of the strength of various correlations will be available soon.

Another field survey was carried out in an office and laboratory building in Karlsruhe during July 2005 in order to compare measurements and subjective votes on thermal comfort in a non air-conditioned indoor environment under German climate conditions (Wagner et al, 2006). Over a period of four weeks, 50 subjects filled in questionnaires twice a day every Tuesday and Thursday, and accompanying measurements were carried out at the workplaces. The actual votes on thermal sensation did not correspond to predicted mean votes, which were calculated with measured data during the interviews, but a very good agreement was seen with adaptive comfort models. Clothing values were noticed to be different to the standard values.

CONCLUSIONS

The issue of thermal comfort at workplaces has gained much importance since the early 1990s. The paper has shown that post-occupancy evaluation in field studies has had a major influence on new understandings, already leading to modified standards (ASHRAE) and new proposals for standards in Europe. These standards, introducing adaptive comfort models, are particularly important for buildings relying on passive cooling strategies in order to meet increasing requirements for energy efficiency (for example, the European Energy Performance of Buildings Directive of 2001). Since the very hot summer of 2003 in Europe at least, upper temperature limits for thermal comfort have been widely discussed (Pfafferott et al, 2004). Unlike existing comfort standards that are derived from laboratory experiments, the adaptive models are derived from field studies and have proved that thermal comfort in non air-conditioned buildings depends more on external climate conditions than on the expectations of building occupants.

Field studies also reveal that the occupants' perceptions of the indoor environment are influenced by numerous parameters. These include building- and workspace-related parameters as well as contextual variables. In contrast to laboratory experiments, boundary conditions cannot be controlled in field studies so there still remain many open questions regarding the 'definition' of the key parameters for overall satisfaction with the workplace, well-being and productivity. Furthermore, little is known about the impact of interrelations between these parameters.

It is interesting that recent experiments in climate chambers have started to implement more 'real building characteristics' in order to converge with 'field experiment facilities'. The well-known climate chambers of the International Centre for Indoor Environment and Energy in Copenhagen, for example, are expanding and have been renamed as 'indoor environment chambers' with the purpose of doing research on air quality, thermal comfort, health and productivity under a wide variety of indoor environment conditions and moderate building energy consumption (Toftum et al, 2004).

The climate chamber of the CBE in Berkeley is a standard office room equipped with all features to control the indoor climate. Nevertheless, many relevant factors that influence perception at workplaces are still not captured, such as typical noises, normal work routines, social surroundings and so on.

One of the major shortcomings of the current status of buildings surveys is a lack of transparency and compliance of methodologies applied for performing the studies and analysing results. Therefore it is easy to understand why the ASHRAE scale, and sometimes the Bedford scale, is in use for most thermal comfort evaluations – despite the fact that field studies for comfort perception or more specific purposes might need different voting procedures. Restricted access to questionnaires, databases or to descriptions of survey procedures and data analysis, prevents a thorough comparison of different studies and their results. A particular problem is the translation of questionnaires into different languages. In the European SCATs project all questionnaires were translated back into English (original language) after having been translated into the languages of the participating countries. This allowed researchers to verify that all details had been interpreted correctly. Cultural influences that give different meanings to the same word in different languages add to this problem.

Methodologies reveal that items like health (Bischof et al, 2003), comfort and productivity are not interpreted in similar ways in the different studies. Further, scales are interpreted differently and the temporal assignment of measurements and votes, for example for perceived indoor temperature, remains unclear in some cases. Another problem is the number of buildings chosen for a particular study. When only one building is studied the survey is not representative, but when a high number of buildings are studied, the number of data to be gathered and analysed increases rapidly. There is a danger that the original intention of field studies to analyse interrelations between different parameters influencing overall satisfaction is being abandoned because the vast amount of data cannot be analysed thoroughly.

The aim of our recent study was therefore to choose a representative sample of buildings and occupant samples, to be able to analyse the overall dependencies on general satisfaction with the workplace and the weighting of its influencing factors (Gossauer et al, 2006). Not all of them do have the same weight in terms of impact on satisfaction, even though this is often assumed. A more standardized data analysis procedure should be followed in the future in order to find the most relevant parameters for comfort, well-being, productivity and health. Important aspects are:

● ratings between different buildings and countries;
● social-cultural, workspace or building-related influences on the ratings;
● groups of buildings with similar characteristics, for example to be found by cluster analysis; and
● sample sizes, gender or seasonal influences on thermal comfort and other comfort parameters.

For further post-occupancy evaluation it is important to have consistent and worldwide available methodologies with validated question modules (such as from the CBE) and with detailed instructions. Methods applied so far for assessing comfort or productivity in running buildings should be critically reviewed for the purpose of modifying existing standards. A high response rate is another important criterion for selection. The studies showed a wide range of response rates, for example, depending on how the questionnaires were distributed. Accessible databases should be provided (like the ASHRAE database), which are of high value for future scientific research.

In a next step, post-occupancy evaluation could be integrated into facility management of commercial buildings to enable more straightforward building performance assessments. Surveys can be used as a diagnostic tool for buildings if they are easy and quick to handle for occupants and give quick and understandable feedback to building managers. The amount of time needed for filling in questionnaires is probably the most crucial issue for gaining a broad acceptance by building managers because it is directly related to the specific economic performance of a person. However, personnel costs usually form the bulk of company expenditure and so should justify the expense of improving the well-being of occupants in their workplace. Appropriate tools could be developed as derivates of research tools. A good example is the CBE survey that asks only for detailed answers to questions about the causes of dissatisfaction.

AUTHOR CONTACT DETAILS

Elke Gossauer, Building Physics and Technical Building Services, University of Karlsruhe, Englerstr. 7, 76131 Karlsruhe, Germany
Email: info@fbta.uni-karlsruhe.de
Andreas Wagner, Building Physics and Technical Building Services, University of Karlsruhe, Englerstr. 7, 76131 Karlsruhe, Germany
Email: info@fbta.uni-karlsruhe.de

REFERENCES AND FURTHER READING

ASHRAE (American Society of Heating, Refrigerating and Air-conditioning Engineers) (2004) 'ASHRAE 55 Standard 55 – Thermal environmental conditions for human occupancy', ASHRAE Inc., Atlanta, GA

Bedford, T. (1936) 'The warmth factor in comfort at work', Rep. Industr. Hlth. Res. Bd., No. 76, London

Bischof, W., Bullinger-Naber, M., Kruppa, B., Hans Müller, B. and Schwab, R. (2003) Expositionen und gesundheitsschädliche Beeinträchtigungen in Bürogebäuden, Ergebnisse des ProKlimA-Projektes, Fraunhofer IBR Verlag, Stuttgart

Bordass, B. and Leaman, A. (2004) 'Closing the loop – post-occupancy evaluation: Next steps – PROBE: How it happened, what it found, and did it get us anywhere?', Proceedings of Cumberland Lodge Conference, Windsor

Bordass, B. and Leaman, A. (2005) 'Making feedback and post-occupancy evaluation routine 1: A portfolio of feedback techniques', Building Research and Information, vol 33, no 4, pp347–352

Brager, G. S. and de Dear, R. (1997) Developing an Adaptive Model of Thermal Comfort and Preference, final report, ASHRAE RP-884, Berkeley, CA

Brager, G. S. and de Dear, R. (1998) 'Thermal adaptation in the built environment: A literature review', Energy and Buildings, vol 27, pp83–96

Brager, G. S. and de Dear, R. J. (2003) 'Historical and cultural influences on comfort expectations', in Cole, R. and Lorsch, R. (eds) *Buildings, Culture and Environment: Informing Local and Global Practices*, Section II, Chapter 11, Blackwell, London

Brager, G. S., Paliaga, G. and de Dear, R. (2004) 'Operable windows, personal control and occupant comfort', *ASHRAE Transactions*, vol 110, Part 2, pp17–35

BREEAM, www.breeam.org, accessed June 2006

Canter, D., Stringer, P. et al (1975) *Environmental Interaction*, Surrey UP, London

Chappells, H. and Stove, E. (2005) 'Debating the future of comfort: environmental sustainability, energy consumption and the indoor environment', *Building Research & Information*, vol 33, no 1, pp32–40

Cooper, G. (1982) 'Comfort theory and practice: Barriers to the conservation of energy by building occupants', *Applied Energy*, vol 11, pp243–288

Cooper, I. (2001) 'Post-occupancy evaluation: Where are you?', *Building Research and Information*, vol 29, no 2, pp158–163

de Dear, R. (2004) 'Thermal comfort in practice', *Indoor Air*, vol 14, no 7, pp32–39

de Dear, R., Brager, G. S. and Cooper, D. (1997) 'Developing, an adaptive model of thermal comfort and preference, *Final Report,* ASHRAE, RP-884

Eriksson, N. (1996) 'Psychological factors and the "sick building syndrome". A case-referent study', *Proceedings of Indoor Air*, vol 6, no 2, pp101–110

Fang, L., Wyon, D. P., Clausen, G. and Fanger, P. O. (2002) 'Sick building syndrome symptoms and performance in a field study at different levels of temperature and humidity', *Proceedings of Indoor Air*, Monterey, vol 3, pp466–471

Fanger, P. O. (1970) *Thermal Comfort. Analysis and Applications in Environmental Engineering*, McGraw-Hill, New York

Fanger, P. O. (1988) 'A comfort equation for indoor air quality and ventilation', *Proceedings of Healthy Buildings*, vol 1, pp39–51

Gagge, A. P., Forbelets, A. P. and Berglund, L. G. (1986) 'A standard predictive index of human response to the thermal environment', *ASHRAE Transactions*, vol 92, no 2b, pp709–731

Gossauer, E., Leonhart, R. and Wagner, A. (2006) 'Nutzerzufriedenheit am Arbeitsplatz– eine Untersuchung in sechzehn Bürogebäuden', submitted for publication

Hellwig, R. T. (2005) *Thermische Behaglichkeit – Unterschiede zwischen frei und mechanisch belüfteten Gebäuden aus Nutzersicht*, Dissertation, University of Munich, Munich

HOPE (2006) http://hope.epfl.ch, accessed June 2006

Humphreys, M. A. (1994) 'Field studies and climate chamber experiments in thermal comfort research', in Oseland, N. and Humphreys, M. (eds) *Thermal Comfort: Past, Present and Future,* Building Research Establishment, Watford, pp52–69

Humphreys, M. A. (2005) 'Quantifying occupant comfort: Are combined indices of the indoor environment practicable?', *Building Research & Information*, vol 33, no 4, pp317–325

Humphreys, M. A. and Nicol, J. F. (2000) 'The effects of measurements and formulation error on thermal comfort indices in the ASHRAE database of field studies', *ASHRAE Transactions*, vol 206, no 2, pp493–502

ISO 7730 (2005) *prEN ISO 7730: Ergonomics of Thermal Environment*, Beuth Verlag, Berlin

Kaczmarczyk, L., Zeng, Q., Melikov, A. and Fanger, P. O. (2002) 'The effect of a personalized ventilation system on perceived air quality and SBS symptoms', *Proceedings of Indoor Air*, vol 4, pp1042–1047

Kröling, P. (1985) *Gesundheits- und Befindlichkeitsstörungen in klimatisierten Gebäuden. Vergleichende Untersuchung zum 'building illness' Syndrom*, Zuckerschwerdt, Munich

Leaman, A. and Bordass, B. (1999) 'Productivity in buildings: The 'killer' variables', *Building Research and Information*, vol 27, no 1, pp4–19

Leaman, A. and Bordass, B (2001) 'Assessing building performance in use 4: The PROBE occupant surveys and their implications', *Building Research and Information*, vol 29, no 2, pp129–143

McCartney, K. J. and Nicol, J. F. (2002) 'Developing an adaptive control algorithm for Europe', *Energy and Buildings*, vol 34, pp623–635

Mendell, M. J. (1993) 'Non-specific symptoms in office workers: A review and summary of the epidemiologic literature', *Proceedings in Indoor Air*, vol 3, no 4, pp227–236

Nicol, J. F. and Humphreys, M. A. (1973) 'Thermal comfort as part of a self-regulating system', *Building Research and Practice (J. CIB)*, vol 6, no 3, pp191–197

Nicol, J. F. and Humphreys, M. A. (2002) 'Adaptive thermal comfort and sustainable thermal standards for buildings', *Energy and Buildings*, vol 34, pp563–572

Nicol, J. F. and Kessler, M. (1998) 'Perception of comfort in relation to weather and adaptive opportunities', *ASHRAE Technical Data Bulletin*, vol 104, no 1, pp1005–1017

Nicol, J. F. and McCartney, K. (2000) *Smart Controls and Thermal Comfort Project*, SCATs final report, Oxford Brookes University, Oxford

Nicol, J. F. and Roaf, S. (2005) 'Post occupancy evaluation and field studies of thermal comfort', *Building Research and Information*, vol 33, no 4, pp338–346

Nikolopoulou, M. and Steems, K. (2003) 'Thermal comfort and psychological adaption as a guide for designing urban spaces', *Energy and Buildings*, vol 35, pp95–101

Pfafferott, J., Herkel, S. and Wagner, A. (2004) 'Müssen unsere Bürogebäude klimatisiert werden?', *HLH – Lüftung/ Klima, Heizung/ Sanitär, Gebäudetechnik*, vol 3, pp24–30

PROBE (1999) *PROBE Strategic Conclusion*, Final Report 4, Center for the Built Environment, Berkeley, CA

Roulet, C. A., Flourentzou, F., Foradini, F., Bluyssen, P. H., Cox, C. H. and Aizlewood, C. (2006a) 'Multi-criteria analysis of health, comfort and energy efficiency of buildings', *Building Research and Information*, vol 34, no 5, pp475–482

Roulet, C. A., Johner, N., Foradini, F., Bluyssen, P. H., Cox, C. H., Oliveria Fernandes, E., Müller, B. and Aizlewood, C. (2006b) 'Perceived health and comfort in European buildings in relation with energy use and other building characteristics', *Building Research and Information*, vol 34, no 5, pp467–474

Rowe, D. (1998) 'Occupant interaction with a mixed media thermal climate control system improves comfort and saves energy', *AIRAH Meeting*, Sydney

Thommen, J. P. (2002) *Management & Organisation, Konzepte, Instrumente, Umsetzung*, Versus Verlag, Zürich

Toftum, J., Langkilde, G. and Fanger, P. O. (2004) 'New indoor environment chambers and field experiment offices for research on human comfort, health and productivity at moderate energy expenditure', *Energy and Buildings*, vol 36, pp889–903

van der Linden, A. C. and Boerstra, A. (2006) 'Adaptive temperature limits: A new guideline in the Netherlands. A new approach for the assessment of building performance with respect to thermal indoor climate', *Energy and Buildings*, vol 38, pp8–17

Voss, K., Herkel, S., Löhnert, G., Wagner, A. and Wambsganβ, M. (2005) *Bürogebäude mit Zukunft – Konzepte, Erfahrungen, Analysen*, TÜV-Verlag, Köln

Wagner, A., Moosmann, C., Gropp, T. and Gossauer, E. (2006) 'Thermal comfort under summer climate conditions: Results from a survey in an office building in Karlsruhe, Germany', *Proceedings of Windsor Conference on Thermal Comfort*, Windsor

Wargocki, P., Wyon, D. P., Sundell, J., Clausen, G. and Fanger, P .O. (2000) 'The effects of outdoor supply rate in an office on perceived air quality, sick building syndrome (SBS) symptoms and productivity', *Indoor Air*, vol 10, no 4, pp222–236

Wilson, S. and Hedge, A. (1997) *The Office Environment Survey: London*, Building Use Studies, London

Zagreus, L., Huizenga, C., Arens, E. and Lehrer, E. D. (2004) 'Listening to occupants: A Web-based indoor environmental quality survey', *Indoor Air*, vol 14, no 8, pp65–74

8

High Dynamic Range Imaging and its Application in Building Research

Axel Jacobs

Abstract
This article describes the theory and application of high dynamic range imaging (HDRI). HDRI is a recent technology allowing the capture of images with a much extended dynamic range whose values represent real-world luminance rather than just arbitrary pixel values. This article lays out the steps necessary to create HDR images, and highlights recent developments in the technology and its applications for building research.

■ *Keywords* – high dynamic range; luminance; digital camera; glare

INTRODUCTION

Advances in digital camera technology have made traditional wet-film cameras all but obsolete. Consumer digital cameras with resolutions approaching that of film cameras have become affordable and enjoy enormous popularity among hobbyists and professionals alike. This success is mostly due to the instant availability of the images, which also allows for sharing them through the internet or other digital media. The photographs may be inspected and even modified by image editors before submitting them to online print centres or simply printing them at home. Digital cameras, however, suffer one major problem that makes it impossible for them to capture scenes as we see them. While the human visual system can adapt to a dynamic range of up to 10,000:1 for parts of a scene and over 12–14 order of magnitude in total, cameras have a much lower dynamic range of typically less than 1000:1. As a result, digital images suffer from lack of detail in the shaded regions and overexposed highlights. The pixel values are not correlated to luminance in the real scene, they merely indicate if one object is more or less bright than another.

   Recent advances in HDRI have shown how those limitations may be overcome. With HDRI, images of the real world may be accurately captured, stored and displayed. This article reviews some of the technologies used to achieve this goal (Battiato et al, 2003;

Reinhard et al, 2006). Apart from having an artistic value, the pixel values of HDR images directly describe the luminance of the objects they depict. With this information and some additional calibration, it is possible to assess the physiological impact of the luminance environment.

CAPTURE

Real scenes encountered in our visual environment usually have large dynamic ranges that may be in excess of 100,000:1 between the brightest and the darkest areas. Digital cameras, however, are only able to capture a rather small part of this. This is due to the properties of the photo sensor within the camera that convert photons to electrons. Due to thermal noise within the active areas, as well as other limitations, a signal will be generated even if no light is incident on the sensor. This dark value limits the lowest irradiance that can be captured by the chip before the usable signal drowns in noise. On the other side of the scale, the sensor can only cope with a certain amount of light before becoming saturated. This not only limits the ability of the one pixel, but the charge may also spill over and affect its neighbours, resulting in the 'blooming' of highlights. Figure 8.1 sketches out the step necessary to convert scene luminance to digital values in a photograph.

These two effects restrict the usable dynamic range of the capturing device. As a result, shadows will be entirely black in the image, making it impossible to distinguish any objects in shady areas, while alternatively, highlights above a certain luminance will be overexposed, making them appear just white. An example indoor scene with high contrasts exhibiting those problems is shown in Figure 8.2.

The solution to this problem is to capture the scene with different exposures (Robertson et al, 1999). Many cameras offer an option called auto-exposure bracketing (AEB), typically allowing the capture of five images in intervals of one exposure value (EV). This will create a sequence of two underexposed photographs (–2, –1EV), one auto-exposure image (±0EV), and two overexposed ones (+1, +2EV). With a combination of manual exposure correction (MEC) and AEB, this range may be extended to ±4EV, while full manual operating mode is limited only by the camera's firmware. For example, the fastest shutter speed on the Nikon 995 is 1/2000s, the slowest one is 8s. Above this, the

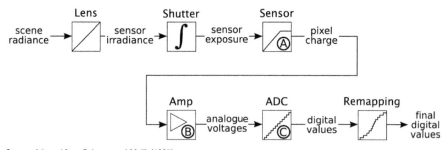

Source: Adapted from Debevec and Malik (1997)

FIGURE 8.1 The image acquisition pipeline

Note: This is an interior scene with very high contrasts, and an auto-exposed image. Note the lack of detail in the dark areas and the washed out highlights.

Source: Axel Jacobs

FIGURE 8.2 Igreja Nossa Senhora de Fatima, Lisbon

camera may be operated in bulb mode, which opens the shutter when the release is pressed until the release is disengaged. Figure 8.3 is an exposure-bracketed sequence taken in full manual mode.

A major inconvenience with the built-in AEB feature is often the fact that for each of the typically five exposures, the shutter has to be triggered. For situations like this it is possible to connect the camera to an external control device, for example the DigiSnap by Harbor (www.harbortronics.com/products2000.asp) that can be set to take a number of images at once. If more control over the camera is required, or if the AEB feature will not cover the dynamic range of the scene, the only solution for a normal consumer digital camera is to control it through a computer or PDA (personal digital assitant) that may be connected via a USB (universal serial bus) or serial lead, or a wireless interface. A software development kit (SDK) is available, for instance from Canon, allowing developers to design remote control software for most of Canon's cameras. An example of one such software is Automated High Dynamic Range Imaging Acquisition (ADHRIA (www2.cs.uh.edu/~somalley/hdri_images.html; O'Malley, 2006). Figure 8.4 is a schematic showing how the original high scene contrast range may be recomposed from low dynamic range (LDR) photographs into an HDR image.

Other camera manufacturers might not offer such development kits, or their models might not have remote control enabled. This is true, for instance, for the Nikon CoolPix

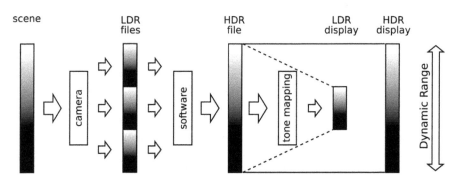

Note: The shortest shown exposure time is 1/60s (top left), the longest 15s (bottom right). Aperture fixed at 2.6, ISO 100.
Taken with Nikon CoolPix 995.
Source: Axel Jacobs

FIGURE 8.3 Sequence of exposure-bracketed images, separated by one f-stop

Note: A number of differently exposed LDR images may be combined to reconstruct the whole contrast range in a HDR image.
Source: Axel Jacobs

FIGURE 8.4 Inability of a single LDR photograph to capture the huge dynamic range of luminance in our environment

range of digital consumer cameras. While the professional cameras of the D series can be remotely controlled, the CoolPix range only offers full remote control over aperture and shutter speeds with older models.

One of the few hand-held HDR cameras is available from TechnoTeam (www.technoteam.de) and developed in collaboration with Rollei. The camera does not require connection to a computer system when images are taken. Instead, the supplied PC software will combine the bracketed images, applying the factory calibration files along the way. The custom firmware takes one image with an automatic exposure and an additional four with $\pm 1/9$th and $\pm 1/3$rd of the shutter speed.

Apart from utilizing consumer cameras and for taking exposure-bracketed sequences, camera systems are also available that require control by a computer system. Two such systems are discussed below.

EXPOSURE MERGING

CAMERA RESPONSE

Once a sequence of exposure-bracketed images has been taken, some post-processing is required. The individual images need to be merged, depending on their exposure, into a HDR image (Grossberg and Nayar, 2003a; 2003b; Shaque and Shah, 2004). While it might seem like a straightforward job for a piece of software to add up pixels, it is actually non-trivial. This is because the value of the pixels in a camera's JPEG file is not linearly proportional to the irradiance on the pixel, but rather processed by the camera to give a response curve that is flatter for dark pixels and steeper for more highly exposed pixels. The shape of this curve is generally not available from the camera manufacturer, and may also vary significantly between two cameras of the same make and even model. It therefore needs to be determined by the HDR software.

In a typical imaging system the irradiance (E) on a pixel is related to the scene luminance (L) as follows:

$$E = L \frac{\pi}{4} \frac{d^2}{h} \cos^4 \varphi \tag{1}$$

where, h is the focal length, d is the diameter of the aperture and φ is the angle between the optical ray and the optical axis (Mitsunaga and Nayar, 1999). In an ideal system, the intensity recorded by the detector is directly proportional to the exposure time.

$$I \approx Et \tag{2}$$

so that

$$I \approx L\,k\,e \tag{3}$$

with

$$k = \cos^4 \varphi/h^2 \quad \text{and} \quad e = \pi(d^2/4)t \tag{4}$$

e may be described as the exposure that can be altered by varying the aperture or the exposure time. While charge coupled devices (CCDs) are designed to have a linear response over a wide range of exposures, the analogue circuitry may introduce non-linearities. Manufacturers also deliberately apply a kind of gamma-mapping to the image to make it more pleasing to the human eye. With M being the final brightness value as stored in the image, knowing the mapping function $g(I)$ allowing us to relate pixel value to image brightness.

$$M = g(I) \tag{5}$$

For this, the inverse mapping function g^{-1} where $I = f(M)$ needs to be determined (Mitsunaga and Nayar, 1999).

Mann and Picard (1995) were the first to point out the problem of response curve calibration and to show a possible solution. They used the function

$$M = \alpha + \beta I^{\gamma} \tag{6}$$

to model the response curve g. The parameter α describes the dark reading and is estimated by taking an image with the lens covered. The scale factor β is set to an arbitrary value. A regression analysis is applied to determine γ. Although their work is an important milestone on the quest for radiometric calibration of camera systems, Mann and Picard's model does not lead to accurate quantitative results.

Debevec and Malik (1997) propose a method to approximate the camera's response with a logarithmic function

$$\ln f(M_{p;q}) = \ln E_p + \ln T_q \tag{7}$$

with p being the spatial index over the pixels, and q being the index over exposures T. The only restriction is that the exact exposures be known. This model works fairly well for images that are not too noisy.

To overcome the limitations of Debevec and Malik's approach caused by the assumption that the response curve can be described with a logarithmic model, Mitsunaga and Nayar (1999) developed a polynomial approach to the problem. Their approach is also more forgiving if the image exposure is not known exactly, which is often the case with inexpensive cameras because the repeatability of the aperture setting is limited.

$$I = f(M) = \sum_{n=0}^{N} c_n M_n \tag{8}$$

The calibration must then determine the minimum order N of the polynomial that is required, as well as the actual coefficients c_n. Response curves for several consumer digital cameras are plotted in Figure 8.5.

The exposure values of the LDR images need to be known for the HDR stitching software to work. They are proportional to the square of the aperture, the equivalent of the ISO rating of the film sensitivity, and inversely proportional to the exposure time with

$$EV = \log_2 \left(\frac{f^2}{T} \frac{S}{100} \right) \tag{9}$$

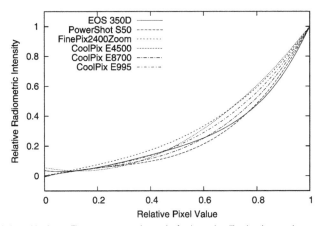

Note: Only the red channel is shown. The curves are not the result of a thorough calibration, but merely examples from WebHDR.
Source: Axel Jacobs; data from WebHDR (http://luminance.londonmet.ac.uk/webhdr)

FIGURE 8.5 Polynomial response curves for selected digital cameras

f being the aperture, *T* being the shutter speed, and *S* being the ISO film sensitivity. Fortunately, all this information is stored inside the JPEG files in what is known as the EXIF (exchangeable image file format) header of the image (http://exif.org). This EXIF, which is now standard in all digital cameras, stores not only the exposure information, but all settings and parameters of the image and the camera it was taken with. Like with many standards, however, the interpretation of the EXIF standard is often somewhat loose, so the software has to be smart enough to compensate for this. For example, the aperture setting may be stored as *Aperture* or *FNumber*, while *ExposureTime* could also be *ShutterSpeed*, and *ISO* may be *CCDISOSensitivity* with some models.

IMAGE ALIGNMENT

It is evident that the camera's response curve can only be determined if the images are perfectly aligned. For this it is necessary to mount the camera on a tripod and, if at all possible, use a remote trigger. However, because special HDR modes are not implemented in consumer cameras yet, it is usually necessary to manually adjust the settings directly on the camera. This results in camera shake, unless a very heavy-duty tripod is used, which is not practical for capturing scenes other than in a laboratory environment.

While it is possible to align the images, the most common alignment operators are based on edge detection or mean pixel threshold. Those may change considerably between images of the same scene taken with different exposures. To overcome this problem, Ward (2003) proposes the use of an alignment operator based on the median of the pixel values that is fairly robust against changes in exposures. His algorithm generates a median threshold bitmap. An image pyramid is then generated by reducing the size of the image in both directions with every successive step. Each image is aligned with its right-hand neighbour by a maximum shiftwidth of one pixel. The smallest image therefore contributes the most significant bit of the final offset.

This image registration works very reliably for the majority of cases and can successfully align hand-held image sequences. However, the only operation is vertical or horizontal shift. If other operations are necessary such as correction of perspective or rotation, more sophisticated algorithms such as developed by Candocia (2003) or Kim and Pollefeys (2004) need to be applied.

CALIBRATION
RESPONSE CURVE CALIBRATION

Assembling a number of exposure-bracketed images into an HDRI image involves determining the camera's response function. This automated process is also known as response curve calibration, the result of which is an HDR image. It differs from LDR images in that the pixel information covers a much wider dynamic range. A known response functionin combination with the exposure information of the images should in theory also leadto radiometrically correct pixel information. This process is known as radiometric calibration.

PHOTOMETRIC CALIBRATION

The exposure value of the individual LDR image, as determined from the EXIF information of the image, is not always exact for two reasons. First, the aperture is adjusted through a diaphragm inside the lens that has only limited repeatability, introducing an error to the aperture number. Second, the exposure time is not necessarily what is set on the camera. The shutter speed is normally changed in such a way that it is halved with each step as the exposure becomes shorter, however, the value is usually rounded off, for example, 1/4, 1/8, 1/15, 1/30, 1/60. With digital cameras, it is more likely to be 1/4, 1/8, 1/16, 1/32, 1/64 and so on, contrary to what the setting might suggest.

Those errors add up, as a result of which the EV cannot be known very accurately. While this is enough to combine LDR images into a usable HDR image, additional photometric calibration must be carried out to obtain per pixel luminance readings of high confidence, as shown in Figure 8.6.

Calibrated luminance cameras come with a set of calibration files that are the result of laboratory tests of the individual units. For improving the accuracy of HDR photographs, the first and easiest procedure is to compare the HDR luminance with readings taken in the real scene with a spot luminance meter. The ratio of the two may be used as a calibration factor for successive images. Using a Minolta LS-100 luminance meter, the author carried out a rather crude photometric calibration for a Nikon CoolPix 990 and two CoolPix 995 cameras. The resulting calibration factors obtained were 1.11, 1.13 and 1.29 as an average over three readings for a scene under an overcast sky.

Inanici (2006) documents a very thorough evaluation of HDR photography, measuring the relative errors for a large number of coloured and grey targets under different light sources. He used a Nikon CoolPix 5400 with a FC-E9 fisheye lens and derived the

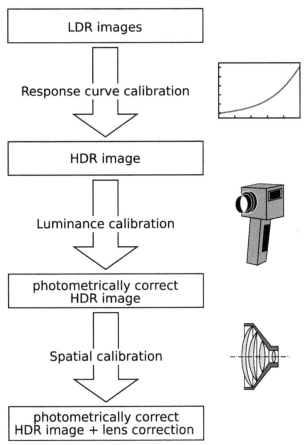

Source: Axel Jacobs

FIGURE 8.6 The information that can be gained from HDR images and its accuracy depend on the level of calibration

following calibration factors: incandescent –0.87, metal halide –0.86, overcast –1.9, and clear sky –2.09.

Anaokar and Moeck (2005) used achromatic Munsell cards to validate the accuracy of HDR reflectance measurements with various consumer grade digital cameras. Their results demonstrate clearly that cameras with built-in thermal noise suppression performed much better than those without. Another interesting finding is that the luminance measurement of dark surfaces with a reflectance of less than 20 per cent resulted in very large errors. They suggest spectral responsivity calibration to overcome this problem. The authors also looked at relative errors in reflectance measurement for different hues of red, green, blue and purple with low and high chroma. Saturated greens and blues were found to produce the largest errors of up to –80 per cent. The most reliable results could be obtained for warm colours with high Munsell values, while

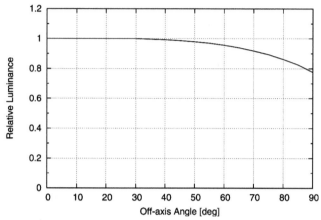

Source: Inanici (2006)

FIGURE 8.7 Reduction of image brightness of a fisheye lens image away from the optical axis

high- to medium-reflectance blues and greens were shown to produce reliable readings down to Munsell value N5. For building interiors, the results should not be too problematic since saturated and dark hues are not frequently found in building materials.

To undertake a more thorough luminance calibration, the vignetting effect has to be taken into account, especially for wide-angle lenses. This was also done by Inanici (2006). His results show that vignetting is virtually zero up to ±30° from the optical axis. At the edges of the image, the luminance is reduced by 23 per cent (Goldman and Chen, 2005). See Figure 8.7 for a plot.

Another systematic error that needs to be corrected is the point spread function (PSF). It is an inherent property of optical systems causing the light passing through the lens to be scattered. This has the effect that the image of an object is somewhat spread out, effecting neighbouring pixels as well. The PSF may be determined computationally by summing up all hot pixels, that is, pixels above a certain threshold in a separate grey scale image (Reinhard et al, 2006). It is then straightforward to remove the PSF from the HDR image by simply subtracting the PSF multiplied by the value of the hot pixel from its neighbours.

SPATIAL CALIBRATION

With a photometric calibration of the camera system, exact photometric information on object luminances may be extracted from the HDR image. If a physiological evaluation of the luminous environment is also desirable, for example, glare evaluation or visibility studies, additional spacial or photogrammetrical calibration is required. This process was first described by Wolf et al (1995) and Berutto and Fontoynont (1995). The objective is to determine for each pixel not only the exact object luminance, but also its position with regards to the camera and the solid angle subtended. With this additional information, physiological scene properties such as unified glare rating (UGR) glare indices may be derived. A prototype UGR meter is shown below.

STORAGE

Images from digital cameras are usually stored in 24-bit JPEG format (www.jpeg.org) with eight bits each for the red, green and blue channels. This results in a resolution of $2^8 = 256$ discrete values for each channel or $2^8 \times 2^8 \times 2^8 = 16$ million values in total. This is also what computer displays can differentiate. Although 16 million colours sounds like a lot, it actually isn't, considering that each channel may only display 256 discrete values. The use of the RAW image format for image capture, which is usually 12 bit or $2^{12} = 4096$ discrete values per channel, may increase the number of discrete values, but to represent the entire dynamic range the human visual system can adapt to, a different file format is needed.

A number of file formats have been developed to overcome the dynamic range limitations of the classical JPEG format as output by digital cameras. They all differ in the dynamic range that they are able to describe, as well as the resulting file size.

RADIANCE RGBE

The most commonly used HDR image format was developed by Greg Ward for the RADIANCE synthetic rendering system (Ward, 1994). Each pixel is represented by a mantissa for each channel and a common exponent. All are stored as 8-bit floating point values, adding up to 32 bits per pixel. The dynamic range of the RGBE format is 76 orders of magnitude (Ward, 1990).

PORTABLE FLOATMAP

The Portable FloatMap HDRI format, which is similar to a floating-point TIFF, assigns 4 bytes to each of the channels. Of those four bytes, one bit is for the sign, one byte for the exponent, and the rest are assigned to the mantissa.

OPENEXR

Developed by Industrial Light & Magic, the OpenEXR file format is now used for all of their film production. It offers a higher dynamic range and colour precision compared to existing 8-bit or 10-bit image formats. The format has support for 16-bit floating-point, 32-bit floating-point, and 32-bit integer pixels. Multiple lossless image compression algorithms are available, some of which can achieve a 2:1 lossless compression result (www.openexr.com).

PFS

Although not a file format as such, the PFS format is mentioned here because it introduces an interesting concept for converting HDR file formats that is rather similar to the netpbm tools for LDR images (http://netpbm.sourceforge.net). Given a number of N different HDR file formats, a total of N^2 converters would be necessary to allow any format to be converted into any other. With an intermediate format, this number is reduced to $2N$. The PFS stream is designed to be this intermediate format. For this, it needs to be generic

TABLE 8.1 Comparison of image size and quality for various file formats

FORMAT	SIZE	%	QUALITY
RAW image data	4.5 MB	100	Full data
RADIANCE XYZE	1.3 MB	29	Perceptually lossless
8-bit BMP	1.2 MB	27	Loss of HDR
8-bit JPEG, 90% Quality	65 kB	1.4	Loss of quality, loss of HDR
JPEG-HDR, 90% Quality	70 kB	1.5	Loss of quality, retention of HDR

Source: JPEG-HDR White Paper (www.brightsidetech.com)

enough to hold a variety of data. The PFS format is not meant to be stored on disk or transferred via slow networks (www.mpi-sb.mpg.de/resources/pfstools/).

JPEG-HDR

It will be apparent that with 32 bits per pixel, an uncompressed HDR image quickly becomes excessively large as image resolution increases. Another problem is that HDR formats are not commonly supported by image editors and viewers. To address this problem, BrightSide developed an HDR extension to the JPEG file format (Ward and Simmons, 2005; Kabaja, 2005).

Table 8.1 shows the reduction in image size and resulting image quality for a number of different file formats for a given scene.

TONE-MAPPING

Current display technology has a very limited dynamic range, so if an HDR image is to be displayed, the dynamic range of the image will have to be compressed to that of the display device. This process is known as tone-mapping.

According to Reinhard et al (2006), tone-mapping operators can be divided into four categories:

● global operators – the same (non-linear) curve is applied to all pixels;
● local operators – the adaptation level is derived for each pixel individually considering the local neighbourhood;
● frequency domain operators – the dynamic range is reduced based on the spatial frequency of the image region; and
● gradient domain operators – a derivative of the image is modified.

A number of operators exists that try to mimic the human visual system. Human vision is a fairly complex process with a highly non-linear response. Tone-mapping operators require the image to be in real-world units, that is, cd/m^2. Because the colour-sensitive cones work for high illumination, with the rods being responsible in low-light environments, one of the effects of human tone-mapping operators is the desaturation of colours for low luminances. Similarly, very bright regions of the image will result in a ceiling luminance thus causing glare (Krawczyk et al, 2005).

The tone-mapping operators briefly introduced below are merely examples chosen because they are implemented in PFSTools (www.mpi-sb.mpg.de/resources/pfstools/).

Many others exist, and the optimum operator for a particular application and output device might require some experimentation from the user (see Reinhard et al (2006) for more exhaustive discussion). Figure 8.8 is the image of the church again, with different tone-mapping operators applied.

GLOBAL OPERATORS

The simplest possible tone-mapping is linear mapping, but because most display devices exhibit a non-linear response, this will result in very dark images if the HDR image has a wide dynamic range. Gamma-corrected linear mapping attempts to correct this problem. Figure 8.9 shows an example. Exponential and logarithmic corrections to linear mapping are also possible.

Drago et al (2003a; 2003b) extend the logarithmic response curves in order to handle a wider dynamic range. A logarithmic compression is applied to the image luminance. The base of the logarithm is varied between 2 and 10, based on the brightness of regions within the image. This results in a preservation of contrast in darker regions and a higher compression for bright regions.

Reinhard et al (2002) took inspiration for their tone-mapping operator from techniques known in traditional wet-film photography. To get the visually best print from a negative, it is not enough to just match contrast and brightness. Scene content, image medium, and

Source: Axel Jacobs

FIGURE 8.8 Linearly tone-mapped image with gamma correction of the church scene displayed with pfsview

Note: Top left to right are auto-exposed LDR image as comparison.
Source: Top left to right: Ashikhmin (2002); Drago et al (2003a); bottom left to right: Durand and Dorsey (2002); Fattal et al (2002); Reinhard et al (2002)

FIGURE 8.9 Results of different tone-mapping operators

viewing conditions must often be considered. It was Ansel Adams who developed the so-called zone system in the 1940s. Now, 60 years later, it is still used successfully because it combines quantitative measurements with artistic image content.

LOCAL OPERATORS

The Pattanaik multi-scale observer operator attempts to model all steps within the human visual system currently known well enough to be modelled (Pattanaik et al, 2000). It is the most complete framework to date, although not strictly speaking necessary for only reducing the dynamic range.

In contrast to the Pattanaik model, Ashikhmin's operator only implements those aspects relevant to dynamic range compression. As a result, the Ashikhmin operator is significantly faster (Ashikhmin, 2002).

FREQUENCY DOMAIN OPERATORS

It can be argued that an image may be thought of as being composed from a LDR component with a high spatial frequency, and a HDR component for low frequencies. If the two can be separated, only one of the components has to be compressed. This concept has been implemented by Durand and Dorsey (2002). It is based on bilateral filtering, which is an edge-preserving operation that blurs the image while keeping edges intact.

GRADIENT DOMAIN OPERATORS

While high-frequency components within an image cause rapid changes in luminance and high gradients, these are much smoother with low-frequency objects. The gradient field, that is, a representation of the pixel changes, is compressed with operators such as Fattal et al's gradient domain compression (2002). By varying the amount of compression, the amount of detail that is preserved in the displayed image may be adjusted.

DISPLAY

Traditional output devices offer only a limited dynamic range. The maximum contrast of paper prints is fairly low, being only about 100:1. Modern TFT flat panel screens achieve a dynamic range of 300:1. This may be up to 5000:1 for digital video projectors. To make the most of HDR images and display them so neither the dark regions nor the highlights are clipped, new display technology is required. What is claimed to be 'the world's first HDR display' was developed by BrightSide. Their model DR-37P is a 37" high definition HDR display, shown in Figure 8.10 (Seetzen et al, 2004).

The display achieves a maximum luminance of around 3000cd/m². Combined with the black value of only 0.015cd/m², the overall contrast ratio is as high as 200,000:1. The unit achieves such high luminance and contrast values through the combination of a liquid crystal display (LCD) panel that is not, like conventional flat-panel monitors, uniformly illuminated by cold-cathode fluorescent lamps, but by an array of 1380 light-emitting diodes (LEDs).

The display supports high definition resolution of 1920 x 1080 pixels. Although there seems to be a discrepancy in numbers between the 1380 LEDs and the roughly 2,000,000 pixels of the high definition display. This apparent mismatch is, however, easily explained. While the human visual apparatus can deal with a luminous range of about 1:1000 in any one scene, this range is dramatically reduced for objects that are very close to one

Source: BrightSide (www.brightsidetech.com)

FIGURE 8.10 DR 37P HDR display by BrightSide

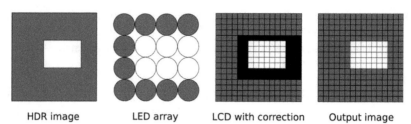

| HDR image | LED array | LCD with correction | Output image |

Source: BrightSide (www.brightsidetech.com)

FIGURE 8.11 The output of the BrightSide display is a combination of the LED array (low spatial frequency) and the LCD modulation (high spatial frequency)

another. If there is a fairly dark object close to a bright light source, we will perceive the dark object as black. If the same dark object was put against a background that is less bright, different shades of dark can actually be differentiated.

It is this effect that the DR 37P exploits. The luminance of the screen is the product of the intensity of the LED and the transmittance of the LCD pixels, as demonstrated in Figure 8.11. Each are controllable with an 8-bit resolution between 'on', 'off', 'fully

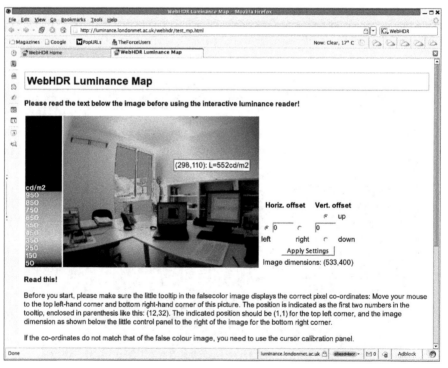

Source: WebHDR (http://luminance.londonmet.ac.uk/webhdr)

FIGURE 8.12 Interactive luminance map in a web browser

transmitting' and 'not transmitting'. In theory, the combined resolution is equivalent to a display with 16 bits per channel or 48 bits in total, however, this is not quite achievable for practical reasons, mostly because the 'off' state of LCDs still allows a not insignificant amount of light to pass through the panel.

APPLICATIONS
LUMINANCE MEASUREMENT
Even without photometric calibration, an HDR image gives a reasonably accurate idea about the luminance distribution in the visual environment and may be used to quickly evaluate scenes such as office environments.

The WebHDR site is very popular with the students on the Masters in Architecture, Energy and Sustainability in the Low Energy Architectural Research Unit (LEARN) (www.learn.londonmet.ac.uk). It allows the upload of up to nine exposure-bracketed photographs of a scene, which are then processed into a HDR image with Ward's *hdrgen* (www.anyhere.com). A screenshot is shown in Figure 8.12. Tone-mapping and false colours may be applied, and the interactive luminance map can be used to quantitatively inspect the scene. Our students use it a lot, especially for taking photographs inside their scale models. To improve the results, it is also possible to calibrate the camera and reuse those results.

If web-based use is not required, there are many other software applications that will create, edit and display HDR images. Some are listed on the WebHDR website, which may also be a starting point for those wanting to explore HDR photography further.

UGR MEASUREMENT
A small number of HDR-enabled digital capturing systems is already available on the market. Due to the image processing that needs to be applied, and also because of the use of industrial-grade imaging units requiring a frame grabber card, most of these systems are based on portable computer systems, making the set-up and use more difficult than that of an ordinary digital camera.

However, because of the computer-based set-up, such systems also offer extra flexibility that ordinary cameras cannot deliver. Figure 8.13, for example, shows a motorized, computer-controlled CCD camera (www.tu-ilmenau.de/site/lichttechnik/index.php?id=843). Because both the optical system and the sensor matrix are calibrated, the pixel value may be directly related to object brightness. Additionally, because the spatial characteristics are known from a photogrammetrical calibration, scene properties that are dependent on luminance and object size/location may also be derived. One such application is the measurement of glare, given as UGR value or in any other system (Wolf, 2004).

As a reminder, the UGR glare formula is given as

$$UGR = 8\log \frac{0.25}{L_b} \sum \frac{I^2 \Omega}{p^2} \qquad (10)$$

with L_b being the background luminance, Ω being the solid angle of the glare source and p being a position index.

Source: TUI-LT (www.tu-ilmenau.de/site/lichttechnik/index.php?id=843)

FIGURE 8.13 UGR meter developed by Technical University of Ilmenau

Instead of adding up the glare contribution of every single pixel, the analysis software of the UGR meter combines neighbouring bright pixels into regions, with each region having an average luminance, solid angle, direction and position index. The UGR formula (or any other glare formula that might be applied through the modular system) is then applied to the resulting glare regions.

With photometrically and spatially calibrated HDR images, the exact luminance, direction and solid angle of each pixel is well defined. Only with this knowledge is it possible to derive values such as UGR glare ratings. Other data such as the illuminance are also obtainable. This is implemented in the UGR introduced above for the glare illuminance, that is, the illuminance at the camera for all or individual glare sources (Wolf, 2004). Figure 8.14 is a screenshot of the UGR meter's control software.

Moeck and Anaokar (2006) extend this concept further. With templates painted in an image processing program, different parts of an HDR were blacked out. This made it possible to derive the illuminance contribution of different objects to the overall illuminance. It was then possible to look at individual features such as windows, artificial light sources or ceiling.

HDR PANORAMA

Another interesting system is the SpheronCam HDR (www.spheron.com) that allows for the capture of HDR panoramic (360°) scenes. It offers a dynamic range of 26 f-stops, and can output the panoramas in a number of popular HDR image file formats. The system is depicted in Figure 8.15. Once set up, the motorized camera can complete an HDR panorama in a few minutes.

Source: TUI-LT (www.tu-ilmenau.de/site/lichttechnik/index.php?id=843)

FIGURE 8.14 Screenshot of the software of the UGR meter

Source: Spheron (www.spheron.com)

FIGURE 8.15 Panoramic HDR capturing system

Source: Spheron (www.spheron.com)

FIGURE 8.16 HDR panorama of Air Zermatt

No manual stitching or image alignment is required with the Spheron system, all this is done with the software provided. Figure 8.16 is a sample panorama taken with the SpheroCam HDR.

For those that don't invest in a SpheroCam HDR, Ward's Photosphere software is capable of assembling HDR panoramas from individual HDR images. It is freely available from Anyhere (www.anyhere.com), but unfortunately only runs under Mac OS X.

OUTLOOK
SENSOR IMPROVEMENT

HDR is still a fairly specialized field that, although showing great potential, has a long way to go before becoming generally accepted and used by consumers. The main reason for this is that taking HDR images is still relatively complicated and fairly time consuming when compared to 'ordinary' photographs.

In this last section we take a closer look at what happens inside a digital camera and how advances in manufacturing and design of photosensitive chips have the potential to radically change the way digital cameras work, thus enabling them to directly output HDR images without requiring any post-processing.

Part of Figure 8.1 that shows the photo sensor inside the digital camera is given again in Figure 8.17. Conversions taking place on the path from irradiance to digital values are marked with A, B and C.

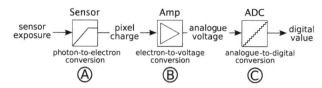

Source: Adapted from Debevec and Malik (1997)

FIGURE 8.17 The conversion of light to a digital value in a digital camera

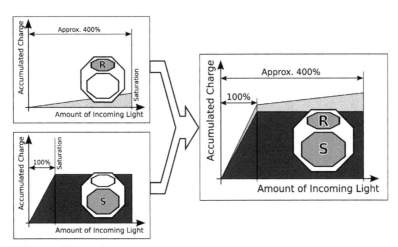

Source: FujiFilm (www.fujifilm.com)

FIGURE 8.18 Extended dynamic range of the Fujifilm SuperCCD sensor

Although it does not qualify as a HDR camera and is only marketed as 'wide dynamic range' Fujifilm's S3 Pro digital camera may be seen as one of the first consumer cameras in this direction. The matrix of CCDs inside the S3 Pro features two types of pixels: S-pixels that do most of the work, and R-pixels that have a much lower sensitivity due to their smaller size. The R-pixels are only used for very bright parts of the image and prevent the highlights from becoming 'blown out'. Fujifilm claim that this approach increases the dynamic range by 400 per cent, as shown in Figure 8.18.

Internally, the S3 Pro utilizes an array of CCDs, which is a proven and reliable technology. A major drawback with CCD chips is their relatively slow read-out due to the sequential nature of the process.

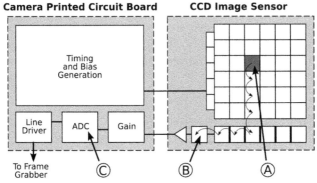

Note: Pixels are read out sequentially.
Source: Litwiller (2001)

FIGURE 8.19 CCD sensor technology

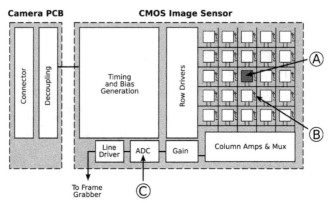

Note: Pixel charge is read out individually, resulting in much faster frame rates.
Source: Litwiller (2001)

FIGURE 8.20 CMOS imaging sensor

Much effort is being undertaken to develop a new breed of imaging sensors based on complementary metal oxide semiconductor (CMOS) technology. The uptake, however, has been relatively slow as a number of problems with this new technology had to be overcome. Products have arrived but are for now restricted to the more expensive range of professional digital SLR cameras. Figures 8.19 and 8.20 show the major components in typical CCD and CMOS imaging chips, while Table 8.2 compares CCD with CMOS in several respects.

CMOS technology differs from CCDs in a number of interesting ways that could potentially lead to a new breed of HDR-enabled cameras (Janesick, 2002; Litwiller, 2005).

With regard to HDRI, the most interesting property of the new breed of CMOS sensors is the fact that they can be read out non-destructively. Combined with the on-chip electronics, this lends itself to the capture of HDR images. Starting with a short exposure time, an automated sequence of images may be taken with increasingly longer exposure times. The on-chip electronics may then combine a HDR image based on the pixel value. Pixels that are too close to the dark value or overexposed would be discarded, while others would contribute towards an exposure-corrected mean value. Due to the integrated on-chip electronics, CMOS sensors are fast enough to even

TABLE 8.2 Comparison of CCD and CMOS imaging chips

	CCD	CMOS
Speed	Slow due to sequential read-out	Superior due to on-chip electronics
Windowing (read-out only portions of the sensor)	Limited	Possible
Anti-blooming	Susceptible to blooming	Immunity to blooming
Energy consumption	Relatively high	Low

take HDR videos at full frame rates and resolutions (Wandell et al, 1999; Lim and EL Gamal, 2001).

FUTURE APPLICATIONS
Sky capture
International meteorological stations monitor weather with automated instrumentation all year round in many locations worldwide. While parameters such as diffuse, global and direct irradiation and illumination are fairly straightforward to determine, sky luminance distributions require relatively costly sky scanners that are made up of a computer-controlled pan-and-tilt head and luminance meter. This set-up is limited in its spatial resolution, and a scan typically consists of 145 directions of 11° angles.

HDR images of the sky hemisphere taken with a 180° fisheye lens provide a much finer resolution that is limited only by the number of pixels on the camera sensors. Sophisticated software may then be used to derive parameters such as percentage of cloud cover or correlated colour temperatures. In particular, statistics on cloud cover are of interest to planners and architects interested in optimizing a building's daylight performance. Predominantly sunny skies and predominantly overcast skies require different approaches to get the most from naturally and freely available daylight (Roy et al, 2001; Stumpfl et al, 2004).

Glare assessment
While the merits and advantages of one glare rating over another are still being debated among lighting experts and physiologists, it is an open secret that daylight glare is virtually impossible to characterize satisfactorily. The daylight glare index (DGI) has been the subject of much criticism and is considered by many as practically useless.

The EU-funded project ECCO-Build tried to remedy this situation by deriving a new index, called daylight glare probability (DGP). It is a combination of existing glare indices and an empirical approach based on the vertical illuminance at the eye, the glare source luminance, its solid angle and position index (Wienold and Christofferson, 2005). The measurement of those parameters are based on HDR fisheye photographs. A new software, *evalglare*, that plugs into the RADIANCE synthetic imaging system was developed to produce DGP values and to visualize the results. Preliminary results show a very high statistical significance and much better repeatability compared to other glare ratings. It is hoped that the DGP, after additional validation and with a larger number of data sets, will become the standard method of assessing glare from windows.

CONCLUSIONS
We have introduced the concept of HDRI. HDR pictures may be produced with any consumer digital camera from a sequence of exposure-bracketed photographs. An HDR image stores values corresponding to the luminance of the real scene with a reasonable accuracy, which may be further improved through additional camera calibration. Although taking HDR images with consumer cameras is time consuming, the process may be eased with custom HDR systems that are beginning to be available on the market. Applications of HDRI, particularly within the building sector, have been demonstrated.

AUTHOR CONTACT DETAILS

Axel Jacobs, Low Energy Architecture Research Unit, LEARN, London Metropolitan University, UK
Email: a.jacobs@londonmet.ac.uk

REFERENCES

Anaokar, S. and Moeck, M. (2005) 'Validation of high dynamic range imaging to luminance measurement', *Leukos*, vol 2, no 2, pp133–144

Ashikhmin, M. (2002) 'A tone mapping algorithm for high contrast images', *Proceedings of 13th Eurographics Workshop on Rendering*, Pisa, Italy

Battiato, S., Castorina, A. and Manucuso, M. (2003) 'High dynamic range imaging for digital still camera: An overview', *Journal of Electronic Imaging*, vol 12, no 3, pp459–469

Berutto, V. and Fontoynont, M. (1995) 'Applications of CCD cameras to lighting research: Review and extension to the measurement of glare indices', *Proceedings of 23rd Session of the CIE*, New Delhi, vol 1, pp192–195

Candocia, F. M. (2003) 'Simultaneous homographic and comparametric alignment of multiple exposure-adjusted pictures of the same scene', *IEEE Transactions on Image Processing*, vol 12, pp1485–1494

Debevec, P. E. and Malik, J. (1997) Recovering high dynamic range radiance maps from photographs', *Proceedings Siggraph*, pp369–378

Drago, F., Myszkowski, K., Annen, T. and Chiba, N. (2003a) 'Adaptive logarithmic mapping for displaying high contrast scenes', *Eurographics*, vol 22, no 3, pp419–426

Drago, F., Martens, W., Myszkowski, K., Chiba, N., Rogowitz, B. and Pappas, T. (2003b) 'Design of a tone mapping operator for high dynamic range images based upon psychophysical evaluation and preference mapping', *Proceedings Human Vision and Electronic Imaging VIII, (HVEI-03)*, pp321–331

Durand, F. and Dorsey, J. (2002) 'Fast bilateral filtering for the display of high-dynamic-range images', *ACM Transactions on Graphics, Proceedings of the 29th Annual Conference on Computer Graphics and Interactive Techniques*, San Antonio, TX, pp257–266

Fattal, R., Lischinski, D. and Werman, M. (2002) 'Gradient domain high dynamic range compression', *ACM Siggraph, Proceedings of the 29th Annual Conference on Computer Graphics and Interactive Techniques*, San Antonio, TX, pp249–256

Goldman, D. B. and Chen, J.-H. (2005) 'Vignette and exposure calibration and compensation', *Proceedings of ICCV '05*, Beijing, China

Grossberg, M. D. and Nayar, S. K. (2003a) 'Determining the camera response from images: What is knowable?', *IEEE Transaction on Pattern Analysis and Machine Intelligence*, vol 25, no 11, pp1455–1467

Grossberg, M. D. and Nayar, S. K. (2003) 'What is the space of camera response functions?', *Proceedings of the 2003 IEEE Computer Society Conference on Computer Vision and Pattern Recognition (CVPR'03)*, Madison, WI

Inanici, M. N. (2006) 'Evaluation of high dynamic range photography as a luminance data acquisition system', *Lighting Research and Technology*, vol 38, no 2, pp123–134

Janesick, J. (2002) 'Dueling detectors' *Spie's OE magazine*, February

Johnson, G. M. and Fairchild, M. D. (2003) 'Rendering HDR images', *IS&T/SID 11th Color Imaging Conference*, Scottsdale

Kabaja. K. (2005) 'Storing of high dynamic range images in JPEG/JFIF files', *CESCG, Central European Seminar on Computer Graphics*, Vienna, Austria

Kim, S. J. and Pollefeys, M. (2004) 'Radiometric self-alignment of image sequences', *CVPR, IEEE Proceedings Computer Vision and Pattern Recognition*, Washington, DC, pp645–651

Krawczyk, G., Myszkowski, K. and Seidel, H.-P. (2005) 'Lightness perception in tone reproduction for high dynamic range images', *Eurographics*, vol 24, no 3, pp635–645

Lim, S. and El Gamal, A. (2001) 'Integration of image capture and processing: Beyond single chip digital camera'', *Electronic Imaging Conference*, San Jose, CA, pp219–226

Litwiller, D. (2001) 'CCD vs. CMOS: Facts and fiction', *Photonics Spectra*, January

Litwiller, D. (2005) 'CMOS vs. CCD: Maturing technologies, maturing markets', *Photonics Spectra*, August

Mann, S. and Picard, R. W. (1995) 'On being "undigital" with digital cameras: Extending dynamic range by combining differently exposed pictures', *Proceedings of IS&T 48th Annual Conference*, Society for Imaging Science and Technology Annual Conference, Washington, DC, pp422–428

Mitsunaga, T. and Nayar, S. K. (1999) 'Radiometric self calibration', *CVPR, IEEE Proceedings Computer Vision and Pattern Recognition*, Fort Collins, CO, vol 2, pp374–380

Moeck, M. and Anaokar, S. (2006) 'Illuminance analysis from HDR images', *Leukos*, vol 2, no 3, pp211–228

O'Malley, S. M. (2006) 'A simple, effective system for automated capture of high dynamic range images', *Proceedings of the Fourth IEEE International Conference on Computer Vision Systems* (*ICVS 2006*), New York, p15

Pattanaik, S. N., Tumblin, J., Yee, H. and Greenberg, D. P. (2000) 'Time-dependent visual adaptation for realistic image display', *Proceedings of ACM SIGGRAPH,* vol 34, pp47–53

Reinhard, E., Stark, M., Shirley, P. and Ferwerda, J. (2002) 'Photographic tone reproduction for digital images', *ACM Transactions on Graphics*, vol 21, no 3, pp267–276

Reinhard, E., Ward, G., Pattanaik, S. and Debevec, P. (2006) *High Dynamic Range Imaging: Acquisition, Display and Image-based Lighting*, Morgan Kaufmann Series in Computer Graphics, Morgan Kaufmann Publishers, San Francisco, CA

Robertson, M. A., Borman, S. and Stevenson, R. L. (1999) 'Dynamic range improvement through multiple exposures', *IEEE International Conference on Image Processing*, Kobe, Japan, 23–27 October

Roy, G., Hayman, S. and Julian, W (2001) 'Sky analysis from CCD images: Cloud cover', *Lighting Research and Technology*, vol 33, no 4, pp211–221

Seetzen, H., Heidrich, W., Stuerzlinger, W., Ward, G., Whitehead, L., Trentacoste, M., Ghosh, A. and Vorozcovs, A. (2004) 'High dynamic range display systems', *Siggraph*, vol 23, no 3, pp760–768

Shaque, K. and Shah, M. (2004) 'Estimation of the radiometric response functions of a color camera from differently illuminated images', *International Conference on Image Processing*, Singapore, 24–27 October, pp2339–2968

Stumpfel, J., Jones, A., Wenger, A., Tchou, C., Hawkins, T. and Debevec, P. (2004) 'Direct HDR capture of the sun and sky', *Proceedings of the 3rd International Conference on Computer Graphics, Virtual Reality, Visualisation and Interaction in Africa (AFRIGRAPH)*, pp145–149

Wandell, B., Catrysse, P., DiCarlo, J., Yang, D. and Gamal, A. (1999) 'Multiple capture single image architecture with a CMOS sensor', *Proceedings of the International Symposium on Multispectral Imaging and Color Reproduction for Digital Archives*, Society of Multispectral Imaging of Japan, Chiba, Japan, pp11–17

Ward, G. [credited as Ward Larson, G.] (1990) 'Radiance file formats', http://radsite.lbl.gov/radiance/refer/refman.pdf

Ward, G. (1994) 'Real pixels', *Graphics Gems*, vol IV, pp80–83

Ward, G. (2003) 'Fast, robust image registration for compositing high dynamic range photographs from handheld exposures', *Journal of Graphics Tools*, vol 8, no 2, pp17–30

Ward, G. and Simmons, M. (2005) 'JPEG-HDR: A backwards-compatible, high dynamic range extension to JPEG', *13th Color Imaging Conference*, Boston, MA

Wienold, J. and Christofferson, J. (2005) 'Towards a new daylight glare rating', Lux Europa 2005, Berlin, pp157–161

Wolf, S. (2004) *Entwicklung und Aufbau eines Leuchtdichte-Analysators zur Messung von Blendungskennzahlen,* Publikationsreihe des Fachgebietes Lichttechnik der TU Ilmenau Nr. 7, Der Andere Verlag, Osnabrück

Wolf, S., Stefanov, E. and Riemann, M. (1995) 'Image resolved measurement of luminance using a CCD Camera', *Light and Engineering,* vol 3, no 3, pp34–44

9

Use of Satellite Remote Sensing in Support of Urban Heat Island Studies

M. Stathopoulou and C. Cartalis

Abstract

Since the early 1960s, numerous satellite sensors have been launched into orbit to observe and monitor the Earth and its environment. Over the years technologies have significantly improved and satellite missions have increased; more importantly perceptions have changed in terms of the potential and usability of satellite remote sensing. One of the application areas of satellite remote sensing is the study of urban areas. This review provides information about the potential of satellite remote sensing for the study of urban climatology, and in particular the surface urban heat island phenomenon.

■ *Keywords* – satellite remote sensing; urban climate; heat island

INTRODUCTION

Urban studies conventionally use demographic or geographic information so as to depict the state of the urban environment, the characteristics of the urban landscape, the changes that have taken place in the course of time and the pressures on the urban area. Such information needs to be spatially and temporally dense, a fact that implies the need for monitoring networks or special experiments that carry the capacity for sufficiently reflecting the area under investigation. Despite improvements in our understanding regarding the significance of urban areas, few cities or urban agglomerations operate monitoring networks that may be considered sufficient. To many urban planners remote sensing is providing a reliable alternative to conventional techniques. Yet considerable discussion is taking place on the real capacity of remote sensing to support urban studies. Such discussion mostly relates to the resolution (temporal, spatial, spectral) of satellite images. If the resolution is low, studies may reflect inaccuracies, especially in the area of land use/cover changes. On the contrary, if the resolution is high, the load of exploitable information highly exceeds – spatially – the respective load of ground data.

It should be mentioned, however, that the temporal and spatial resolutions of a satellite image are anti-correlated, meaning that the better the spatial resolution, the worse the temporal one, and vice versa. When discussing satellite images, spectral resolution also needs to be explored. Although the study of many urban applications is based on images in the visible and near infrared parts of the spectrum, satellite images in the thermal infrared have an excellent potential in terms of supporting urban microclimatic studies.

A clear example is studies of the surface urban heat island (SUHI) phenomenon, namely where warm urban temperatures at the surface contrast with those of surrounding, non-urban areas. Other studies examine canopy layer heat islands (CLHIs) and boundary layer heat islands (BLHIs). Both are examined on the basis of measurements of air temperatures acquired with the use of thermometers. SUHIs are strong both day and night due to the warmer urban surfaces and also depend on a number of factors such as weather (particularly wind and clouds), geographic location (mainly climate and topography), time of day and season, city form (materials, surface characteristics of the city, thermal properties, building dimensions) and city functions (emissions in the atmosphere, heat from energy usage).

A large number of scientists (some of whose work is discussed in this review) have used remote sensing as a tool for the study of urban heat islands. The capacity of thermal sensors to record heat radiation originating from various surfaces, either in the atmosphere (for example, cloud tops) or on the ground, allows the extraction of surface temperature, a parameter that directly indicates the thermal state of the surface under investigation at the time of measurement. It should be mentioned that the time of the measurement is an important parameter as remote sensing provides a 'snapshot' of the state of the thermal environment, which may be either regarded as an acceptable representation of the overall state of the thermal environment or as a time-confined 'picture' of the thermal environment. The use of remote sensing for the study of the urban heat island (UHI) depends also on the temporal resolution of satellite images. Moreover, considerable effort needs to be invested in our knowledge of the emissivity factors of the materials in the area under investigation. This is not a trivial issue; on the contrary, the accuracy of the extracted results may well depend on the respective accuracy of the used emissivity factors.

SURFACE URBAN HEAT ISLAND STUDIES
USING LOW–MEDIUM RESOLUTION THERMAL SENSORS

The conceptual framework of employing satellite remote sensing data for surface urban heat island assessment was laid out by Rao (1972). Inspired by his work, many SUHI studies were conducted making use of thermal data from early satellite systems (Carlson et al, 1977, 1981; Matson et al, 1978; Price, 1979; Carlson, 1986; Vukovich, 1983; Kidder and Wu, 1987; Balling and Brazel, 1988; Roth et al, 1989; Caselles et al, 1991; Gallo et al, 1993a, 1993b).

Lee (1993) examined the potential of utilizing the NOAA-AVHRR thermal data to study surface urban heat islands. Using the Seoul metropolitan area as an application site, Lee employed AVHRR data at channel 4 (10.5–11.5µm) to map the pattern of brightness temperature distribution in the city. The isotherms of the AVHRR-derived brightness temperatures were projected on a land cover map of Seoul city and its thermal pattern was examined in relation to land use and land cover. From analysis, the warm areas during daytime were identified; these were associated with central business, industrial and densely residential districts. Lee found that the AVHRR-derived brightness temperatures were highly correlated with the near-surface air temperatures ($R^2 = 0.73$) and the ground surface temperatures ($R^2 = 0.72$) and supported the potential of utilizing the AVHRR-derived brightness temperatures to derive the air and ground surface temperature fields in a city, helpful in assessing surface urban heat islands. Finally, Lee defined an empirical relationship between heat island intensity and the population of Korean cities that could be applied for estimating the population size necessary for heat island formation.

Streutker (2002) studied the SUHI of Houston, Texas by the use of night-time NOAA-AVHRR thermal data. Surface temperature maps of the city were produced for 21 selected dates covering a two-year period (1998–1999). The magnitude and spatial extent of the SUHI were defined for every satellite image in the data set by applying a Gaussian method. More specifically, the AVHRR-derived surface temperature image data were fitted to a two-dimensional Gaussian surface superimposed on a planar rural background, which was accomplished by performing a least-squares fit to the natural logarithm of the surface temperature. Streutker found the surface urban heat island intensity of Houston to range from 1.06–4.25°C depending on season and weather conditions, whereas the spatial extent of the SUHI varied, with longitudinal extent between 20 and 70km, and latitudinal extent between 15 and 30km. According to Streutker, the Gaussian method of SUHI magnitude calculation proved to be more accurate than alternative methods proposed in literature because this method provided a substantially lower heat island magnitude as it was not influenced by localized high surface temperatures within the urban web. It was also found that although the heat island magnitude and the rural temperature showed a weak correlation to the SUHI spatial extent, the SUHI magnitude was inversely correlated with the rural temperature.

Dousset and Gourmelon (2003) studied the summertime microclimate of the Los Angeles and Paris metropolitan areas by combining satellite multi-sensor data with *in-situ* air quality data in a geographic information system (GIS) platform. More specifically, they used a series of summertime NOAA-AVHRR thermal data corresponding to various times of day (morning, afternoon, evening, night) in order to produce images of average surface temperatures from which statistics (maximum, minimum and diurnal amplitude) were extracted. In addition, they used SPOT-HRV visible and near infrared data at a 20m spatial resolution to derive the land cover classification for the cities. Moreover, visible and near infrared data from both sensors were employed to derive the normalized difference vegetation index (NDVI). The average land surface temperature (LST) images of each city were combined with their corresponding land cover image and *in-situ* air quality data in a GIS. In this way, the temporal and spatial variations of LST were investigated with respect to land cover and air quality. The researchers found a strong SUHI of 7°C for Paris at night and a 'negative' surface urban heat island during the day, where commercial/industrial and airport regions as well as satellite suburbs displayed higher surface temperatures than downtown Paris. In addition, diurnal variations in LST were found to correlate with surface ozone concentration from *in-situ* experiment. For the city of Los Angeles, the statistics of the diurnal cycle of LST as derived from the average images revealed a strong heat island in the range of 7.5–9°C during the day and a weaker heat island in the range of 2–5°C at night, which was attributed to the influence of the Pacific Ocean. Finally, for both cities, a negative correlation between LST and NDVI was derived that explained the cooling effect of urban parks.

Stathopoulou et al (2004) studied the SUHIs of four coastal cities in Greece (Thessaloniki, Patra, Volos and Heraklion) with the use of night-time NOAA-AVHRR image data covering a week during summer (14–21 July 1998). The researchers combined the thermal AVHRR data in channels 4 (10.5–11.5μm) and 5 (11.5–12.5μm) with the CORINE Land Cover (CLC) database for Greece to determine land cover-based surface emissivities needed for the estimation of LST. This combination resulted in the production of surface temperature maps of an enhanced spatial resolution of 250m for the cities under study. The heat islands of the cities were investigated and correlated to land cover, whereas urban heat island intensities were measured on pixel-based maximum surface temperature differences between urban and rural sites observed at a weekly scale. SUHI intensities were found to be independent of city size but dependent on the urban development. The study also demonstrated: first, the correlation between surface temperature and land cover, as high surface temperatures were associated with industrial, commercial, densely-built and port areas; and second, the influence of topography on surface temperature. The mean weekly night-time SUHI intensities of the cities were estimated: 8°C for the cities of Thessaloniki and Heraklion, and 7°C for the cities of Patra and Volos.

NOAA-AVHRR data in the thermal infrared have also been used to derive parameters related to urban environment. Stathopoulou et al (2005a) used a series of summer AVHRR thermal data acquired at midday (14:00 UTC) to estimate the discomfort index (DI) in Athens, Greece. More specifically, they used the thermal data to derive the surface temperature at the spatial resolution of 1.1km. Surface temperatures were correlated to *in-situ* near-ground air temperatures obtained from three types of weather stations (urban, suburban and rural) and a strong correlation was found ($R^2 = 0.85$) which was applied to convert AVHRR surface temperatures to air temperatures. Relative humidity was assessed in terms of dew point temperature and of a split-window estimate of atmospheric precipitable water. Comparison between the AVHRR-derived DI values and the *in-situ* station DI values revealed a good agreement ($R^2 = 0.79$), with a RMSE (root mean square error) of 1.2°C and a bias of 0.9°C, thus supporting the potential of using AVHRR thermal data to define the DI spatial variation at a higher resolution (1.1km) than that achieved by meteorological stations.

Cooling degree days (CDDs) are another parameter estimated with the use of AVHRR thermal data. CDDs allow the assessment of the effect ambient air temperature has on the energy performance of buildings. Stathopoulou et al (2006) estimated the CDDs for the metropolitan city of Athens based on the surface temperature excess from a surface base temperature, the latter being determined from an empirical expression that related midday AVHRR surface temperature with daily air temperature as recorded at standard meteorological stations. This approach offered the advantage of the direct application of daily satellite data for the definition of CDDs in urban areas at a spatial resolution of 1.1 km. Knowledge of the spatial distribution of CDDs supports the definition of location specific standards for thermal insulation of buildings so as to ensure satisfactory energy performance.

Hung et al (2006) utilized TERRA-MODIS image data to perform a comparative study of the SUHIs of eight mega-cities in Asia. They used both daytime and night-time MODIS data acquired over the 2001–2003 period to produce surface temperature maps for the eight cities at 1km spatial resolution. In addition to MODIS data, they used high spatial resolution Landsat ETM+ data to derive land cover information and to relate SUHI patterns to surface characteristics. The researchers adopted the Gaussian method proposed by Streutker (2002) to measure the spatial extents and magnitudes of the SUHIs for the eight cities. The diurnal and seasonal patterns of the satellite-derived SUHIs revealed that all cities exhibited significant heat islands. The relationship between heat islands and surface properties as well as the correlation between SUHI magnitude and city population was also determined.

USING HIGH RESOLUTION THERMAL SENSORS

Carnahan and Larson (1990) used Landsat 5 thermal data (10.4–12.5μm) to examine the surface urban heat sink of the Indianapolis metropolitan area, a phenomenon that is termed the 'negative urban heat island', meaning that the urban area is cooler than the surrounding rural area. Employing satellite data, they detected the heat sink, measured its magnitude and analysed the mechanisms favourable to its formation. They found that specific temporal and surface characteristics contributed to the development of the Indianapolis urban heat sink, such as low thermal inertia of the soil.

Kim (1992) investigated the surface urban heat island of the metropolitan area of Washington DC with the use of seasonal Landsat TM data. Initially, a single TM scene was used for classification and mapping of the spatial patterns of five different surface categories within the city: bare soil, grass, urban, forest and river. Following, with the use of short-wave TM spectral bands, the net solar radiation (W/m^2) absorbed by the five surface categories was determined for all images in the data set. It was found that the absorption profile of an urban area could be characterized as a compromise between bare soil and woodland. Combining a one-dimensional (1-D) evapotranspiration model with the thermal TM data, Kim determined the following thermal energies by surface categories and seasons: sensible heat flux, outgoing long-wave radiation and latent heat of evaporation, which were in turn approximated by simple linear regressions of surface reflectance and solar radiation incident. In this way, the energy balance status and the SUHI of Washington DC were examined. It was found that the urban heating in Washington began to appear by mid-morning in summer and could yield a difference of 10°C against nearby forest area due to a large excess in heat (sensible heat fluxes) from the rapidly heating dry urban surfaces.

The applicability of the Landsat TM thermal data (10.4–12.5μm) to microclimate monitoring was also investigated by Nichol (1994). In her study, surface temperature data for nine high-rise housing estates in Singapore were derived from thermal TM data by applying a method that enhanced the spatial resolution of the thermal data from 120m to

30m, according to vegetation status as measured from the reflective TM data. The study confirmed the inverse relationship between the satellite-derived surface temperature patterns and moisture availability as represented by the vegetated surfaces, as well as the correspondence between satellite-derived surface temperatures and near-ground air temperatures during the daytime. The Landsat image data was found to be suitable for identifying the potential heat-generating surfaces and for examining the spatial aspects of the urban microclimate. Finally, recommendations were made regarding the utilization of the image-derived results in urban planning and future landscaping policies for the city of Singapore.

Aniello et al (1995) examined the usefulness of Landsat TM image data and a GIS for mapping the micro-urban heat islands (MUHIs) of the urban area of Dallas, Texas. As defined in their study, MUHIs are isolated urban locations that produce 'hot spots' within a city. They used the visible TM image data for urban land-use classification from which tree cover information was extracted. The thermal TM data (10.4–12.5µm) were used to generate a thermal pattern distribution map of the study area, which was merged with land-use classification in a GIS to show the location of MUHIs and trees. MUHIs that developed within the city of Dallas were estimated to be 5–11°C warmer than surrounding areas by mid-morning and were attributed to lack of tree cover associated with impervious surfaces, such as parking lots, roads, apartment complexes, shopping centres and newly developed residential neighbourhoods that had less tree cover than older neighbourhoods.

Nichol (1996) evaluated Landsat TM-derived thermal data for their correspondence to building geometry and landscape features in two of Singapore's high-rise housing estates. Field data of T_a (air temperature) and T_s (surface temperature) were collected in the highly built-up environment of the housing estates at different levels above ground (1–43m) and at different building orientations for examining the relationship between air and surface temperature. In addition, thermal TM data were used to generate a surface temperature distribution map corresponding to the housing estates. The maps of surface temperature patterns were co-registered and then overlaid to a GIS coverage of streets and buildings for examining the effects of building geometry and landscape to surface temperature. In her study, Nichol attested the close agreement between spatial distributions of satellite and field T_s values as well as field T_s and T_a, thus supporting the capacity of thermal satellite image data for monitoring the daytime thermal environment of an urban area. Moreover, both satellite and field data confirmed the cooling effect of urban tree canopies. In addition, the thermal effects of building geometry were supported by either satellite or field data, as differently oriented street canyons (with respect to solar azimuth) for the same date exhibited different thermal characteristics: irradiated canyons were found to be warmer than shady canyons. As stated in the study, 'the image data are specific enough to indicate the relative temperatures of individual buildings or building complexes' (Nichol, 1996).

Adopting a satellite remote sensing approach, Lo and Quattrochi (2003) studied the impact of land use/cover change in the city of Atlanta in Georgia for a period of 25 years (1973–1998) on SUHI development, environmental quality and health implications. They used 11 Landsat images of the Atlanta metropolitan area (13 urban counties), between

1973 and 1998, to map the land use and land cover of the city giving emphasis to differentiating between urban and non-urban uses. In particular, they applied a six-class classification scheme: 1) high-density urban use, 2) low-density urban use, 3) cultivated/exposed land, 4) cropland or grassland, 5) forest and 6) water. The land use/cover change analysis was performed on the basis of statistics (coverage percentage) extracted from the produced land use/cover maps of the Atlanta metropolitan area. The researchers also investigated the impact of land use/cover change on the land surface characteristics. They used thermal TM image data for surface temperature mapping and reflective TM image data for vegetation mapping in terms of NDVI. Analysis of the surface temperature and vegetation maps revealed the negative relationship between NDVI values and surface temperatures, as well as the impact of land use/cover changes and topography on surface temperature. The relation between surface temperatures and NDVI, as extracted from the Landsat data, to volatile organic compound (VOC) and nitrogen oxides (NO_x) emissions, as well as to the rates of cardiovascular and chronic lower respiratory diseases, were also examined by the researchers. The two biophysical variables (surface temperature and NDVI) showed a strong correlation with the VOC and NO_x emissions through a canonical correlation analysis, suggesting that the satellite-derived biophysical variables would be good predictors of VOC and NO_x emissions.

The impact of urban land cover expansion on SUHI intensity and surface temperature patterns was also examined by Weng and Yang (2004). In particular, they investigated the relationship between urban development and the SUHI in the city of Guangzhou in South China, and discussed the effectiveness of the city's greening campaigns to mitigate the SUHI effect. To better understand the impact of urban development on land surface temperatures, the researchers employed Landsat TM thermal data to obtain the thermal signature of each land cover type. It was found that different land cover types had distinctive thermal responses. The thermal effect of spatial arrangement of different land cover types was also described. As suggested in the study, satellite-derived land use/cover maps can be used by urban planners in conjunction with land surface temperature maps to identify the optimal location, size and shape of a proposed land use/cover type or to model the thermal impact of a proposed land use/cover change, so as to minimize the adverse thermal effects of the urban development. Moreover, it was demonstrated that effective landscape policies are those that maintain the level of greenness in line with the pace of urban development, thus making the urban environment sustainable.

In a later study, Stathopoulou et al (2005b) employed Landsat ETM+ thermal data to define SUHIs in the metropolitan city of Athens and the cities of Thessaloniki, Patra, Volos and Heraklion, which are the four largest cities of Greece. Landsat 7 ETM+ thermal data were used to map the surface temperature of the cities during the warm season (May to September). The satellite-derived surface temperature patterns were analysed and related to urban use and land cover. For each city the following five land cover type categories were defined: 1) urban/densely built area, 2) suburban/medium built area, 3) mixed urban area (mainly including urban uses such as industrial/commercial and transport uses, mines/dump and construction sites), 4) rural area and 5) sea water. Surface emissivity according to land cover/use was determined and incorporated into surface temperature retrieval. Overlaying land cover maps to surface temperature maps in a GIS environment using a zonal summary operation scheme, they computed the mean surface temperature by land cover/use category.

CONCLUSIONS

An overall conclusion is that satellite remote sensing can be considered as a valuable tool for the study of SUHIs. The combined use of satellite data in various spectral regions is also beneficial as the derived urban characteristics (land use/cover, vegetation cover and so on) provide considerable insights with respect to the presence, the intensity and variability of the SUHI. Results can be further improved by the combined use of ground data and can also be used as the starting point for the calculation of such products as the discomfort index or the cooling degree days.

Disadvantages do exist and relate mostly to the spatial and temporal resolutions of satellite images as well as to the knowledge of the properties of the emitting surfaces. In particular studies of SUHIs, which need satellite images of high temporal resolutions (order of hours), correspond to low spatial resolutions (order of 1km), a fact that may be considered highly confining for detailed results. In the event that satellite images of medium resolution are used instead (120–250m), the respective temporal resolution drops considerably (from 2–3 days to 16 days). However, given that SUHI studies focus on a microclimatic phenomenon that in the long run reflects timescales exceeding those of hours or even days, such (temporal and spatial) resolutions may be considered adequate for large urban agglomerations. Problems do arise, however, in smaller urban areas because medium spatial resolutions may lead to only a few pixels of exploitable information. Problems also arise in the examination of the daily variation of SUHIs, as limited information is available from satellite night passes. The authors consider this as the most pronounced drawback in terms of existing satellite sensors/missions.

AUTHOR CONTACT DETAILS

M. Stathopoulou, Remote Sensing and Image Processing Laboratory, Department of Physics, Division of Applied Physics, University of Athens, University Campus, Build. PHYS-V, 157 84 Athens, Greece
C. Cartalis, Remote Sensing and Image Processing Laboratory, Department of Physics, Division of Applied Physics, University of Athens, University Campus, Build. PHYS-V, 157 84 Athens, Greece

ACKNOWLEDGEMENTS

This study is co-funded by the European Social Fund and National Resources – (EPEAEK II) PYTHAGORAS.

REFERENCES

Aniello, C., Morgan, K., Busbey, A. and Newland, L. (1995) 'Mapping micro-urban heat islands using Landsat TM and a GIS', *Computers and Geosciences*, vol 21, no 8, pp965–969

Balling, R. C. and Brazel, S. W., (1988) 'High-resolution surface temperature patterns in a complex Urban Terrain' *Photogrammetric Engineering and Remote Sensing*, vol 54, no 9, pp1289–1293

Carlson, T. N. (1986) 'Regional-scale estimates of surface moisture availability and thermal inertia using remote thermal measurements', *Remote Sensing Review*, vol 1, pp197–247

Carlson, T. N., Augustine, J. A. and Boland, F. E. (1977) 'Potential application of satellite temperature-measurements in analysis of land-use over urban areas', *Bulletin of the American Meteorology Society,* vol 58, pp1301–1303

Carlson, T. N., Dodd, J. K., Benjamin, S. G. and Cooper, J. N. (1981) 'Satellite estimation of the surface energy balance, moisture availability and thermal inertia', *Journal of Applied Meteorology*, vol 20, pp67–87

Carnahan, W. H. and Larson, R. C. (1990) 'An analysis of an urban heat sink', *Remote Sensing of the Environment*, vol 33, pp65–71

Caselles, V., Garcia, M. J., Melia, J. and Cueva, A. J. (1991) 'Analysis of the heat-island effect of the city of Valencia, Spain, through air temperature transects and NOAA satellite data', *Theoretical and Applied Climatology*, vol 43, pp195–203

Dousset, B. and Gourmelon, F. (2003) 'Satellite multi-sensor data analysis of urban surface temperatures and land cover', *ISPRS Journal of Photogrammetry and Remote Sensing*, vol 58, pp43–54

Gallo, K. P., McNab, A. L., Karl, T. R., Brown, J. F., Hood, J. J. and Tarpley, J. D. (1993a) 'The use of a vegetation index for assessment of the urban heat island effect', *International Journal of Remote Sensing*, vol 14, no 11, pp2223–2230

Gallo, K. P., McNab, A. L., Karl, T. R., Brown, J. F., Hood, J. J. and Tarpley, J. D. (1993b) 'The use of NOAA AVHRR data for assessment of the urban heat island effect', *Journal of Applied Meteorology*, vol 32, pp899–908

Hung, T., Uchihama, D., Ochi, S. and Yasuoka, Y. (2006) 'Assessment with satellite data of the urban heat island effects in Asian mega cities', *International Journal of Applied Earth Observation and Geoinformation*, vol 8, pp34–48

Kidder, S. Q. and Wu, H-T. (1987) 'A multispectral study of the St. Louis area under snow-covered conditions using NOAA-7 AVHRR data', *Remote Sensing of Environment*, vol 22, pp159–172

Kim, H. H. (1992) 'Urban heat island', *International Journal of Remote Sensing*, vol 13, no 12, pp2319–2336

Lee, H. (1993) 'An application of NOAA AVHRR thermal data to the study of urban heat islands', *Atmospheric Environment*, vol 27B, pp1–13

Lo, C. P. and Quattrochi, D. A. (2003) 'Land-use and land-cover change, urban heat island phenomenon, and health implications: A remote sensing approach', *Photogrammetric Engineering and Remote Sensing*, vol 69, no 9, pp1053–1063

Matson, M., McClain, E. P., McGinnis, D. F. Jr and Pritchard, J. A. (1978) 'Satellite detection of urban heat islands', *Monthly Weather Review*, vol 106, pp1725–1734

Nichol, J. E. (1994) 'A GIS-based approach to microclimate monitoring in Singapore's high-rise housing estates', *Photogrammetric Engineering and Remote Sensing*, vol 60, no 10, pp1225–1232

Nichol, J. E. (1996) 'High resolution surface temperature patterns related to urban morphology in a tropical city: A satellite-based study', *Journal of Applied Meteorology*, vol 35, no 1, pp135–146

Price, J. C. (1979) 'Assessment of the urban heat island effect through the use of satellite data', *Monthly Weather Review*, vol 107, pp1554–1557

Rao, P. K. (1972) 'Remote sensing of urban "heat islands" from an environmental satellite', *Bulletin of the American Meteorological Society*, vol 53, pp647–648

Roth, M., Oke T. R. and Emery, W. J. (1989) 'Satellite-derived urban heat islands from three coastal cities and the utilization of such data in urban climatology', *International Journal of Remote Sensing*, vol 10, pp1699–1720.

Stathopoulou, M., Cartalis, C. and Keramitsoglou, I. (2004) 'Mapping micro-urban heat islands using NOAA/AVHRR images and CORINE Land Cover: An application to coastal cities of Greece', *International Journal of Remote Sensing*, vol 25, no 12, pp2301–2316

Stathopoulou, M., Cartalis, C., Keramitsoglou, I. and Santamouris, M. (2005a) 'Thermal remote sensing of Thom's Discomfort Index (DI): Comparison with in situ measurements', *Proceedings of SPIE, Volume 5983, Remote Sensing for Environmental Monitoring, GIS Applications and Geology V*, pp59830K1–59830K9.

Stathopoulou, M., Cartalis, C. and Andritsos, A. (2005b) 'Assessing the thermal environment of major cities in Greece', *Proceedings of the 1st International Conference on Passive and Low Energy Cooling for the Built Environment, PALENC 2005*, 19–21 May, Santorini, Greece.

Stathopoulou, M., Cartalis, C. and Chrysoulakis, N. (2006) 'Using midday surface temperature to estimate cooling degree-days from NOAA-AVHRR thermal infrared data: An application for Athens, Greece', *Solar Energy*, vol 80, pp414–422

Streutker, D. R. (2002) 'A remote sensing study of the urban heat island of Houston, Texas', *International Journal of Remote Sensing*, vol 23, no 13, pp2595–2608.

Vukovich, F.M., (1983) 'An analysis of the ground temperature and reflectivity pattern about St. Louis, Missouri, using HCMM satellite data' *Journal of Climate and Applied Meteorology*, vol 22, pp560–571

Weng, Q. and Yang, S. (2004) 'Managing the adverse thermal effects of urban development in a densely populated Chinese city', *Journal of Environmental Management*, vol 70, pp145–156

10

Thermal Behaviour of the Diffusive Building Envelope: State-of-the-Art Review

Kai Qiu, Fariborz Haghighat and Gerard Guarracino

Abstract

This paper presents a state-of-the-art review of the thermal behaviour of the diffusive building envelope, through which slow and nearly uniform air flows. With the airflow through the envelope, it will exchange heat with solid parts, thus changing the thermal performance of the envelope. The fundamental heat exchange in the air cavities of the envelope is first investigated. Next, the study focuses on the airflow and heat transfer in the porous building materials, in order to assess airflow impact on the thermal performance of the dynamic insulation. The influence of moisture, especially possible phase change, is also discussed. The investigation shows that heat exchange in the envelope will have an impact on the energy consumption of the building. Meanwhile, much work is needed concerning engineering implementation; for example, to provide the suitable analytical methods and to avoid possible condensation.

■ *Keywords* – building envelope; heat transfer; airflow; infiltration; dynamic insulation; moisture

INTRODUCTION

The building envelope's role is to separate the indoor environment from the outdoor environment, to provide a safe, comfortable and healthy environment for the building occupants, as well as to reduce the building's energy consumption. Therefore the accurate estimation of heat loss through the building envelope is important for the determination of the building's energy consumption. Current research has shown that there is a significant impact of infiltration airflow on the building envelope's thermal performance because it can behave as a heat exchanger. This means that actual heat losses due to infiltration could be only a fraction of that obtained using conventional methods – conductive heat losses. As most building materials are porous media, there is an interaction between pores and cracks and the solid matrix, thus the wall can act as a heat exchanger that reduces heat loss through the building envelope. This type of building envelope, the dynamic insulated wall, has not been explored in the design of advanced integrated façades.

This paper presents a detailed literature review of the thermal behaviour of the diffusive building envelope. Starting from the fundamentals of heat exchange in the envelope, the thermal performance of the diffusive building envelope is discussed, and the potential implementation of diffusive building envelopes is addressed.

AIR INFILTRATION THROUGH CAVITIES

Bhattacharyya and Claridge (1995) investigated the infiltration heat exchange performance in insulated walls considering airflow through an air cavity. They developed a simplified one-dimensional (1-D) model to examine the general features of the dependence of infiltration heat exchange on path and flow rate, using the measured quantities as key parameters. In the model for a single stud cavity specimen, airflow is through a straight-line crack between the inlet, at the bottom of the stud cavity, and the outlet, which is at a variable height position of the specimen. The steady-state energy conservation equation in the wall is expressed as

$$\dot{m}c_p \frac{dT_f}{dx} + U_{left}\, w(T_f - T_c) = U_{right}\, w(T_h - T_f) \tag{1}$$

where $\dot{m}$ is the mass flow rate through the envelope, C_p is the heat capacity of air, U_{left} is the U-value of the left side of the wall, U_{right} is the U-value of the right side of the wall, w is the width of the wall, T_f is the air temperature, T_c is the cool side (outdoor) temperature and T_h is the hot side (outdoor) temperature, respectively.

The actual infiltration heat load can be calculated after solving equation (2) and obtaining the T_f value. The infiltration heat exchange efficiency (IHEE), η, was defined to represent the extent of heat recovery, namely

$$Q_{inf} = (1 - \eta)\, Q_{inf\,C} \tag{2}$$

where Q_{inf} is the actual heat load due to infiltration and $Q_{inf\,C}$ is the actual heat load due to infiltration calculated by the conventional method.

By considering the infiltration and exfiltration of the whole building envelope simultaneously, the infiltration heat exchange efficiency was derived. Corresponding to this model, Claridge et al (1995) performed a systematic investigation using a number of experimental set-ups. To consider different infiltration paths, they adopted the concept of concentrated flow (CF) and diffuse flow (DF). CF was defined as a short-path flow, such as airflow through doors, windows and large cracks or holes. DF was defined as long-path flow in which the air travels some distance before leaving the wall.

The main drawback of Bhattacharyya and Claridge (1995) is that the 1-D air cavity model straight through the wall does not reflect the physical phenomena and the complex condition of air infiltration. Meanwhile, to use their model, the length of infiltration path L should be measured or assumed in advance.

One obvious advantage of the air cavity heat exchanger model is that it is possible to derive an analytical solution. The recent work of Barhoun and Guarracino (2004) follows this idea, starting from the model of Bhattacharyya and Claridge (1995), which states that the heat transfer coefficient for the left side of the wall U_{left} and for the right side of the wall U_{right} changes with the outlet position in the y direction. They assumed that the air flows vertically throughout the wall from the bottom to the top, and that the airflow path is in the horizontal centre of the wall. Based on this, by solving the linear differential equation, an analytical solution of infiltration temperature, T_{inf}, as a function of the vertical coordinate, y, was derived as follows:

$$T_f(y) = (T_c - \frac{b}{a}) \exp(- ay) + \frac{b}{a} \tag{3}$$

where

$$a = \frac{w}{\dot{m}c_p}\left(U_{left} + U_{right}\right)$$

$$b = \frac{w}{\dot{m}c_p}\left(T_c U_{left} + T_h U_{right}\right) \tag{4}$$

The air cavity approach was also adopted by Chebil et al (2003) to simulate a transient three-dimensional (3-D) heat and air transfer phenomenon in a typical multi-layer wall structure. Instead of solving complex fluid flow equations, the airflow in each cavity was considered as one-dimensional and correlations for the Nusselt number from the literature were used to determine the heat exchange between air cavities and solid parts. In the case of air infiltration, the following Nusselt number corrections for forced convection (Kohonen and Virtanen, 1987) were used:

$$Nu_f = 1.85 \ (Re \cdot Pr \cdot D_h/H)^{1/3} \qquad Re \cdot Pr \cdot D_h > 70$$

$$Nu_f = 7.54 \qquad\qquad\qquad\qquad Re \cdot Pr \cdot D_h < 70 \tag{5}$$

where Re is the Reynolds number, Pr is the prandtl number, D_h is the air cavity hydraulic diameter and H is the wall height.

They studied a number of scenarios for flow paths and airflow rates. Their simulation results showed that the scenario with the longest airflow path has the most obvious overall heat load reduction effect.

The convection between air in the cavity and the building envelope solid matrix influences the overall thermal performance of the envelope. However, for the modelling of infiltration heat exchange, the air cavity models regard that the airflow is CF within the cavity and heat exchange occurs in the boundary layer along the borders of the air cavity. Therefore the effect of heat exchange on the building envelope's thermal performance is not significant, unless a long infiltration path is constructed. Comparatively, heat exchange between the infiltration air and the solid matrix of the porous material is more applicable in terms of combining heat recovery and ventilation is small residential buildings.

MODEL OF AIR INFILTRATION THROUGH POROUS BUILDING MATERIAL

The infiltration air not only flows through small cavities in the envelope, but also through the porous building material. This is especially obvious for the insulation, which has a high porosity. As the air moves through the porous material, it becomes warmer if the outside temperature is cooler than room temperature, and its temperature gradually becomes closer to the room air temperature because of the heat exchange with the solid matrix. This heat exchange phenomenon has been studied by several researchers. Krarti (1994) presented a 1-D model and, assuming air transfer through the wall with a constant flow rate, developed an analytical solution by adopting convective boundary conditions on both inside and outside surfaces, assuming that outdoor and indoor temperature follows a sinusoidal law. An investigation was conducted based on the analytical solution of the model to evaluate the thermal performance of a dynamic insulation system integrated with a whole building. The variation of 'wall efficiency' in a day was obtained according to the calculated thermal load under steady-state and transient boundary conditions. The results show that the wall efficiency changes from 15 per cent to 45 per cent over a day, with an average of 22 per cent.

Buchanan and Sherman (2000) applied computational fluid dynamics (CFD) simulations to better reflect the fundamental physics of the air infiltration heat recovery process with the building envelope. The model integrated the airflow and heat transfer within a room and in the wall. Airflow and energy transport in the room are determined using the Navier-Stokes and energy equations, considering laminar flow. The wall plywood sheathing is represented as an impermeable, solid material and energy transport is calculated by the conduction equation. Insulation is represented as porous media and airflow through it is calculated using Darcy's law. To represent the heat transfer in the porous media, the following effective conductivity was used in the energy equation:

$$k_e = \varepsilon k_a + (1-\varepsilon)k_s \tag{6}$$

where k_e is the effective conductivity, k_a is the air conductivity, k_s is the conductivity of the solid matrix, and ε is the porosity of the material.

To investigate the influence of the airflow path, several kinds of wall configurations, in which air enters and leaves the insulation at different positions, were used in the simulations. Their simulation results show that the infiltration heat recovery process mainly occurs within the wall structure and in the vicinity of the wall structures. Therefore, it is not necessary to study in detail the building interior for the heat exchange phenomenon in the wall. In view of this, Abadie et al (2002) only considered the boundary layer flow near the wall rather than the flow in the whole room. To do this they added an air cavity on both the inner and outer side of the wall to extend the computational domain. These additional spaces are large enough that the boundary layer flows on the wall surface are largely unaffected by the CFD computational space boundaries, and small enough that the increase in CFD computational time is acceptable.

A model containing the flow in the air gap and flow through porous media is also presented by Chen and Liu (2004) to study heat transfer and airflow in a composite solar-wall system, which consists of a glass cover, a porous absorber, an air gap and a massive thermal storage wall. Their work showed that heat from the incident solar radiation can be stored in the porous absorber, and thus the porous absorber can work as a good thermal insulator when sunlight is not available. However, heat exchange performance is not addressed in the model.

The model by Buchanan and Sherman (2000) does not focus on the wall. As the simulation in the wall was coupled with the indoor airflow, it took too long for the solution to converge. The technique of variable time step is therefore needed to speed up the solution for real applications. Furthermore, the models based on the airflow and heat transfer in the porous media do not connect with the potential implementation of the heat exchange in the envelope. As heat exchange actually occurs in the building envelope, investigations need to be focused on the airflow and heat transfer in the envelope.

Besides the numerical simulation, an analytical model was also presented by Buchanan and Sherman (2000), by dividing the whole exterior wall into an infiltration-affected area and a non-affected area. However, how to select the infiltration area is not mentioned, which is not convenient for engineering implementation. Meanwhile, the influence of the infiltration path is not included in the model. Numerical simulation of airflow and heat transfer in the porous material by Qiu (2006) shows that although the air infiltration path (that is, straight through (ST) configuration or low inlet/high outlet (LH) configuration) influences the heat exchange performance, the airflow is almost uniform with a very low velocity along the infiltration path, if the material is homogeneous and isotropic. Therefore, it is possible to derive an analytical model including the airflow path influence. A model is thus presented by dividing the whole wall into ventilation area and non-ventilation area. To decide the ventilation area, which is the infiltration-affected area, parametric study was performed using the numerical model. It was found that the ventilated area is a function of the wall thickness. It is also a function of airflow rate. However, the variation of the ventilated area is much smaller than the variation of the airflow rate. A simplified representation for the ventilation area is hence suggested for the engineering implementation of estimating the actual heat loss by a conduction-infiltration coupled process.

Air can be heated or cooled before it enters the room by exchanging heat with the diffuse building envelope. This means that the building envelope, or part of it, can have airflow through it for the purpose of ventilation and reducing the total energy consumption of the building.

THERMAL PERFORMANCE OF DYNAMIC INSULATION

Dynamic insulation is a combination of conventional insulation and dynamic heat exchange within a wall. Being an efficient way of preheating infiltrated air, it is regarded as one possible method for reducing building envelope heat losses and improving indoor air quality. The existing technology of dynamic insulation can be divided into the following two catalogues: 1) using cavities to circulate the fluid (mostly air) in the wall; and 2) breathing wall.

In the first kind of configuration, the airflow direction in the cavities is generally parallel to the wall height. The interaction of moving air and other parts of the wall makes the wall act as a heat exchanger. The second configuration allows the gas (mostly air) to transfer through the permeable insulation. The interaction of air and solid matrix can also act as a contra-flux mode heat exchanger, where the gas moves through the wall in the opposite direction of the heat flow (Baker, 2003). Though ventilated walls using air cavities have been adopted, such as Baily (1987) and Chebil et al (2003), currently the research and application of dynamic insulation systems focus on the breathing wall.

The dynamic insulated wall component usually consists of three main sub-layers:

- the ventilated external envelope sub-layer;
- the dynamic insulation sub-layer; and
- an air gap separating these two sub-layers.

The external sub-layer could be a prefabricated reinforced concrete slab (Dimoudi et al, 2004), or a profiled metal sheet (Baker, 2003). The ventilation air can be introduced from the bottom or top of the external envelope sub-layer. The dynamic insulation sub-layer may consist of layers of breathing materials, including a compressed strawboard layer, mineral wool and thin paperboard, or cellulose fibre insulation. These breathing materials let the air enter the room under a pressure difference between interior and exterior.

Meanwhile, to assure the perpendicularity of different layers and the strength of the component, the dynamic insulation is positioned on a wooden frame crossed with wooden internal studs. The layout of timber studding is also helpful for minimizing thermal bridging through the structure. At the same time, to assure the uniform airflow and hence 1-D heat transfer through the wall, for the design that permits air to come into the wall from bottom of the external layer, the lower 1.0m of the wall is constructed to have a higher air resistance (Dimoudi et al, 2004).

ANALYTICAL MODEL

Heat transfer processes in the dynamic insulation are nearly one-dimensional. A steady-state 1-D model is presented by Taylor et al (1996):

$$k_e \frac{d^2 T(x)}{dx^2} - u\rho C_{pa} \frac{dT(x)}{dx} = 0 \qquad (7)$$

k_e is the effective thermal conductivity of the wall material, T is the temperature, ρ is the density of the air, u is the air velocity, C_{pa} is the heat capacity of air.

By setting the temperature at two boundaries, being the outdoor temperature and indoor temperature, and assuming air velocity in the wall is constant, an analytical solution was obtained:

$$\frac{T - T_o}{T_i - T_o} = \frac{\exp\left(\frac{u\rho c_p x}{k_e}\right) - 1}{\exp\left(\frac{u\rho c_p L}{k_e}\right) - 1} \qquad (8)$$

where T_i is the temperature at indoor side, T_o is the temperature at outdoor side, and L is the thickness of the dynamic insulation.

The model has been extended to a three-layer wall system and assumed a steady-state condition:

$$q_{c0} = \frac{\ln G}{G - 1} \frac{T_i - T_o}{R} \tag{9}$$

where q_{c0} is the conductive heat flux at outer surface, $R = \sum\limits_{i=1}^{3} \dfrac{L_i - L_{i-1}}{k_i}$ is total thermal resistance of the envelope, and $G = \exp(\mu \rho c_p R)$ is a dimensionless parameter.

According to this 1-D steady-state model, the dynamic U-value was derived as follows (Qiu and Haghighat, 2006)

$$U_{dyn} = \frac{Pe}{R(e^{Pe} - 1)} \tag{10}$$

where $Pe = \dfrac{u \rho_a C_{pa} L}{k_e}$, and $R = \dfrac{L}{k_{eff}}$ is the effective thermal resistance of insulation material in the static condition.

It is clear that the dynamic U-value is a function of the air velocity. It is regarded as a characteristic of the performance of dynamic insulation (Baker, 2003). Fig 10.1 shows the dynamic U-value and corresponding heat exchange efficiency as a function of the airflow rate, for two lengths of the dynamic insulation: 0.1m and 0.2m. The effective conductivity of the insulation is 0.035W/mK. It can be seen that the static condition has the highest

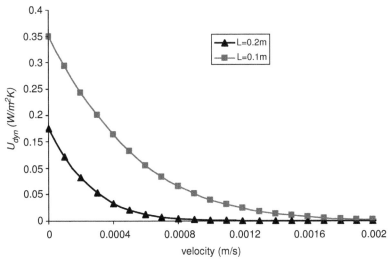

Source: Qiu (2006)

FIGURE 10.1 Dynamic U-value with velocity

heat transfer rate, and this heat transfer rate decreases with the airflow rate. At very low air velocity, that is, below 0.002m/s, the dynamic insulation with a thickness of 0.2m has a lower heat transfer rate than the dynamic insulation of 0.1m. However, when the air velocity is greater than 0.002m/s, the conductive heat loss at the exterior surface of the wall tends to be zero for a wall with a thickness of either 0.1m or 0.2m.

Thus the key principle of dynamic insulation is to determine the insulation thickness and assure the range of airflow rate. On one hand, to supply outdoor fresh air, promote the heat exchange between air and insulation, and decrease the risk of condensation, the airflow rate should not be too slow. On the other hand, if the temperature of incoming air is equal to the outdoor temperature, the convective heat loss increases as the airflow rate increases, hence the overall heat loss will increase.

Affected by a combination of the above factors, the airflow rate needs to be in a well-defined range. This is not reflected by the results in Figure 10.1, as the dynamic U-value decreases with the increase of airflow rate. This is because the result is obtained from an equation (10) in which only the conductive heat loss is included.

A model considering the influence of both conduction and convection was derived by Qiu and Haghighat (2006), which is as follows:

$$U_T = \frac{Pe}{R(e^{pe} - 1)} + \frac{Pe}{R} \tag{11}$$

In the derivation of equation (8), the boundary condition was that the temperature at the outside and inside nodes of the wall are constant and equal to that of outdoor and indoor temperature. However, influenced by the air film near the wall, the temperature at the boundaries may be different from that, especially for the interior surface. As the airflow rate is very slow in the dynamic insulation wall, the more suitable boundary condition for the analysis is as follows (Taylor and Imbabi, 1997):

$$-k_e \frac{dT}{dx}\bigg|_{x=L} = h(T_i - T_{wi}) \tag{12}$$

where T_i is the indoor temperature, T_{wi} is the inside surface temperature and h is the convective heat transfer coefficient.

Under steady-state conditions, if considering the influence of thermal resistance of inside and outside air films, the temperature of the internal surface of the dynamic insulated wall can be calculated as (Taylor and Imbabi, 1997; Gan, 2000)

$$\frac{T_i - T_{si}}{T_i - T_0} = \frac{R_i \exp(Pe)}{R_i \exp(Pe) + \frac{\exp(Pe) - 1}{k_e Pe/L} + R_0} \tag{13}$$

T_{si} is the temperature of the inner surface of the dynamic wall, R_i is the thermal resistance of the interior air film, and R_0 is the thermal resistance of the exterior air film. By performing a simple calculation, it would be seen that the interior surface temperature decreases with the airflow rate.

Corresponding to this theory, results have been obtained in the above investigations concerning the performance of dynamic insulated walls. First, when calculating the interior or exterior surfaces temperatures, the suitable boundary condition is that the conductive heat flux at the wall surface, rather than the net heat flux, is equal to the flux incident on the wall from the environment. However, in the assessment of heat loss through dynamic insulation over the static equivalent, the influence of both interior and exterior surface air films can be neglected. Furthermore, it is shown that the air film influences the interior and exterior surface temperatures, so it impacts on the thermal comfort of occupants in the room.

Concerning the application of dynamic insulation in building design, Krarti (1994) simulated a room with a dynamic insulated wall, applying an analytical model, and showed that the overall energy saving may reach 20 per cent. Also based on the theoretical analysis, the product Energy*flo*™ cell, developed by the Environmental Building Partnership Ltd in the UK, is claimed to reduce the required heating and cooling load by 10 per cent, compared to the Scotland Building Regulation Standard (Environmental Building Partnership Ltd, 2005). As the overall energy saving by dynamic insulation is generally only about 10 per cent, it is less attractive because this in only marginally better than a supply-and-exhaust ventilation system (Morrison and Karagiozis, 1992). One option for this problem is to combine dynamic insulation with natural ventilation, or hybrid ventilation, taking the advantage of pressure gradients created by wind and stack effect. For example, Etheridge and Zhang (1998) reveal that a strong synergy exists between dynamic insulation and wind energy. Therefore there is good potential to extend the application of dynamic insulation to naturally ventilated systems. However, further investigation needs to be carried out on this issue for its implementation.

EXPERIMENTAL INVESTIGATION

Baker (2003) reports the results of measurements of ventilation rate between 14 to 47m³ per hour. The results illustrate that though the heat transmission coefficient at the external surface of the dynamic insulation decreases with ventilation rate, the heat transmission coefficient at the internal surface increases with the airflow rate. At an airflow rate of 40m³ per hour, the two coefficients are 0.1W/m²K and 0.85W/m²K, respectively. The former is lower and the latter is higher than the static U-value, which is equal to 0.26W/m²K. This means that with the increase of the airflow rate, the portion of conduction loss recovered by ventilation increases, however, more energy is required to raise the incoming air temperature to that of the room. This result corresponds with the analytical model prediction, which demonstrates that the generally used expression of dynamic U-value of the building envelope only accounts one part of the real heat loss through the wall.

Considering the heat flux at the internal surface of the dynamic insulation, the experimental data of Baker (2003) is illustrated in Figure 10.2 and compared with the prediction of Qiu and Haghighat (2006) in equation (11). The ventilation heat loss is determined by the average incoming temperature at the interior surface of the wall. We can see that these results are in good agreement. Thus, the analytical model can predict the energy consumption of the dynamic insulated wall with good precision.

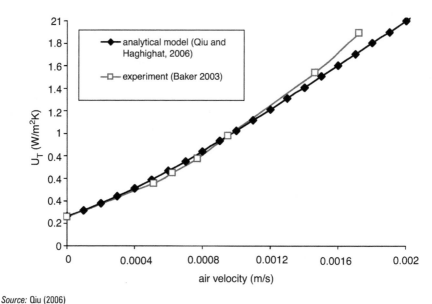

Source: Qiu (2006)

FIGURE 10.2 Comparison of experiment data and analytical model results

The experiment results also show that with the increase of ventilation rate, the internal surface temperature decreases. At the 40m³ per hour airflow rate, for a 15K temperature difference between indoor and outdoor air, the internal surface temperature is 2K lower than that without ventilation. Furthermore, it is demonstrated that for a fixed ventilation rate, the heat recovery efficiency is not a constant value and is strongly correlated with solar radiation. The solar radiation was found to contribute to the heat gains in the incoming infiltration at about 5 per cent over the test period.

The hourly changed internal–external surface temperature difference and conductive heat flux at the internal surface under real outdoor conditions during a one-day period are reported by Dimoudi et al (2004). The results showed that, depending on the ambient conditions during the day, the operation of the dynamic insulation may change from contra-flux mode to pro-flux mode.

The experimental work of Crowther (1995) focused on the influence of interior and exterior air film coefficients on the thermal performance of the dynamic insulation. The results show that the heat resistance of the interior surface is determined by natural convection and radiation, while the heat resistance of the outer surface is almost all contributed by radiation.

INFLUENCE OF MOISTURE

The previous models do not consider the influence of moisture. However, the infiltration/exfiltration air may contain water vapour. When this humid air flows through the envelope, condensation may occur under certain conditions. This will influence the heat exchange performance of the envelope. To include the influence of phase change, investigation needs to be carried out taking into consideration both heat and moisture transfer simultaneously. Heat and moisture transfer in porous material has been investigated since the pioneering work of Philip and DeVries (1957) and Luikov (1966). The energy equation is incorporated with vapour and liquid transport caused by capillary forces. The capillary force is represented in terms of gradients of the moisture content and temperature. Later, Whitaker (1977) averaged the transport equations on a representative elementary volume (REV) at the continuum level and obtained the governing equations in a higher level.

Vasile et al (1998) developed a model concentrating on heat transfer, by attempting to describe the influence of the moisture level on heat transfer occurring through hollow vertical terracotta bricks. According to the authors, the 1-D energy conservation equation including the effects of phase change by condensation/evaporation is as follows:

$$\rho_m c_p \frac{\partial T}{\partial t} = \frac{\partial}{\partial x}\left(k\frac{\partial T}{\partial x}\right) - L_v \varepsilon' \frac{\partial \dot{m}}{\partial x}$$

(14)

where t is time, ρ_m is the density of the dry material, T is temperature, k is the thermal conductivity of the air, L is the water evaporation energy, m is the total mass flux and ε' is the phase change rate. The phase change rate is determined by the variation of heat flux and water contents. The simulation results show that heat flux through hollow bricks can change from 25 to 35 per cent from low to high humidity.

However, it is clear that only conduction heat transfer and phase change is considered, while influence of air infiltration is neglected in this model. The airflow in the envelope generally decreases the possibility of condensation, especially when it is a constant airflow, such as that in the dynamic insulation. However, it is pointed out that under some conditions, such as solar radiation on wet timber cladding, condensation may occur in the dynamic insulated wall (Taylor and Imbabi, 1998). This needs to be further investigated, combined with the investigation of impact of solar radiation.

Corresponding with theoretical work on heat and moisture transfer in the building envelope, several software packages have been developed, such as 1D-HAM (Hagentoft and Blomberg, 2000). The model used in the package 1D-HAM is a basic model that includes heat, air and moisture transfer in a multi-layer porous envelope in one direction. In the simulation model, the airflow rate through the envelope is assumed to be uniform, determined by the constant flow resistance and pressure difference across the envelope. Moisture is transferred by diffusion and convection in the vapour phase and heat is transferred by conduction, convection and latent heat. To simplify the problem, water transport is not included in their model.

However, these software packages focus on the moisture transfer in the building envelope and assessment of the possibility of condensation. Heat transfer is not addressed and the energy consumption of the buildings is not included. Therefore they need to be further developed if they are to be used to analyse the thermal behaviour of the diffusive building envelope.

CONCLUSION

The assessment of the thermal behaviour of the diffusive building envelope needs to consider the heat exchange performance of the envelope. Influenced by heat exchange, the temperature in the envelope deviates from that of pure conduction, and the temperature of infiltration is not equal to the outdoor air temperature upon its entering the room. This will have an impact on the energy consumption of the buildings.

Heat exchange occurs in the boundary layer in the air cavity in the building envelope. Investigations have been carried out to reveal its effect on the overall thermal performance of the envelope. The results illustrate that because of the heat exchange in the envelope, total heat loss through the building envelope is less than that estimated by the conventional method. However, the influence of infiltration recovery between air in the cavity and solid parts is usually not significant, unless the air moves in a specially designed long path.

When infiltration air flows through the porous building material it exchanges heat with the solid matrix of the building envelope. Numerical simulation of airflow and heat transfer in the porous media suggests that, influenced by air infiltration, the building envelope can be divided into a ventilated area and a non-ventilated area, and heat exchange occurs in the ventilated area.

Dynamic insulation is a technique of combining the airflow through the envelope and heat exchange in the porous insulation for the purpose of ventilation and reducing energy consumption. Concerning its thermal performance, heat transfer coefficients based on a 1-D steady-state model are a suitable tool for the engineering implementation.

Condensation is another factor of heat exchange in the building envelope. Though combined heat and moisture transfer in the building envelope has been studied, further investigation needs to be carried out and combined with the detailed condition of the engineering application.

AUTHOR CONTACT DETAILS

Kai Qiu, Department of Building, Civil and Environmental Engineering,
Concordia University, Montreal, Canada
Fariborz Haghighat, Department of Building, Civil and Environmental Engineering,
Concordia University, Montreal, Canada
Gerard Guarracino, Département Génie Civil Bâtiment, Ecole Nationale des Travaux Publics de l'Etat, Lyon, France

ACKNOWLEDGEMENTS

The authors are grateful for the financial support provided by the Natural Science Engineering Research Council of Canada.

REFERENCES

Abadie, M. O., Finlayson, E. U. and Gadgil, A. J. (2002) *Infiltration Heat Recovery in Building Walls: Computational Fluid Dynamics Investigations Results*, Lawrence Berkeley Laboratory Report, LBNL-51324, University of California, CA

Baily, N. R. (1987) 'Dynamic insulation systems and energy conservation in buildings', *ASHRAE Transactions*, vol 93, pt 1, pp447–466

Baker, P. H. (2003) 'The thermal performance of a prototype dynamically insulated wall', *Building Services Engineering Research and Technology*, vol 24, pp25–34

Barhoun, H. and Guarracino, G. (2004) 'Evaluating the energy impact of air infiltration through walls with a coupled heat and mass transfer method', *Proceedings of CIB Conference*, Toronto, Canada

Bhattacharyya, S. and Claridge, D. E. (1995) 'Energy impact of air leakage through insulated wall', *Journal of Solar Engineering*, vol 117, pp167–172

Buchanan, C. R. and Sherman, M. H. (2000) *A Mathematical Model for Infiltration Heat Recovery*, Lawrence Berkeley Laboratory Report, LBL-44294, University of California, CA

Chebil, S., Galanis, N. and Zmeureanu, R. (2003) 'Computer simulation of thermal impact of air infiltration through multilayered exterior walls', *Eighth International IBPSA Conference*, Eindhoven, Netherlands

Chen, W. and Liu, W. (2004). 'Numerical analysis of heat transfer in a composite wall solar-collector system with a porous absorber', *Applied Energy*, vol 78, pp137–149

Claridge, D. E., Liu, M. and Bhattacharyya, S. (1995) 'Impact of air infiltration in frame walls on energy loads: Taking advantage of the interaction between infiltration, solar radiation, and conduction', in Modera, M. P. and Persily, A. K. (eds) *Airflow Performance of Building Envelopes, Components, and Systems, ASTM STP 1255*, American Society for Testing and Materials, Philadelphia, pp178–196

Crowther, D. R. G. (1995) *Health Considerations in House Design*, PhD thesis, University of Cambridge, Cambridge, UK

Dimoudi, A., Androutsopoulus, A. and Lykoulis, S. (2004) 'Experimental work on a linked, dynamic and ventilated, wall component', *Energy and Buildings*, vol 36, pp443–453

Environmental Building Partnership Ltd (2005) *Technical Bulletin 1*, Rev.1. Environmental Building Partnership, Aberdeen, UK

Etheridge, D. W. and Zhang, J. J. (1998) 'Dynamic insulation and natural ventilation: Feasibility study', *Building Services Engineering Research & Technology*, vol 19, pp203–212

Gan, G. (2000) 'Numerical evaluation of thermal comfort in rooms with dynamic insulation', *Building and Environment*, vol 35, pp445–453

Hagentoft, C. E. and Blomberg, T. (2000) '1D-HAM Coupled heat, air and moisture transport in multi-layered wall structures', www.buildingphysics.com/manuals/1dham.pdf, accessed April 2006

Kohonen, R. and Virtanen, M. (1987) 'Thermal coupling leakage flow and heating load of buildings', *ASHRAE Transactions*, vol 93, pp2302–2318

Krarti, M. (1994) 'Effect of airflow on heat transfer in walls', *Journal of Solar Energy Engineering*, vol 116, pp35–42

Luikov, A. W. (1966) *Heat and Mass Transfer in Capillary-Porous Bodies*, Pergamon Press, Oxford

Morrison, I. D. and Karagiozis, A. N. (1992) 'Energy impact of dynamic wall ventilation', *Proceedings of 18th Annual Energy Society of Canada*, Edmonton, Alberta

Philip, J. R. and DeVries, O. A. (1957) 'Moisture movement in porous materials under temperature gradients', *Transactions of American Geophysics Union*, vol 38, pp222–232

Qiu, K. (2006) *Air Infiltration and Heat Exchange Performance of the Building Envelope*, PhD thesis, Concordia University, Montreal

Qiu, K. and Haghighat, F. (2006) 'Simulation of thermal performance of buildings incorporating breathing wall elements', *Proceedings of eSim 2006 Conference*, Toronto, Canada

Taylor, B. J., Cawthorne, D. A. and Imbabi, M. S. (1996) 'Analytical investigation of the steady-state behaviour of dynamic and diffusive building envelopes', *Building and Environment*, vol 31, pp519–525

Taylor, B. J. and Imbabi, M. S. (1997) 'The effect of air film thermal resistance on the behaviour of dynamic insulation', *Building and Environment*, vol 32, pp397–404

Taylor, B. J. and Imbabi, M. S. (1998) 'Application of dynamic insulation in buildings', *Renewable Energy*, vol 15, pp377–382

Vasile, C., Lorente, S. and Perrin, B. (1998) 'Study of convective inside cavities coupled with heat and mass transfer through porous media: Application to vertical hollow bricks – a first approach', *Energy and Building*, vol 28, pp229–235

Whitaker, S. (1997) 'Simultaneous heat, mass and momentum transfer: A theory of drying', *Advanced Heat Transfer*, vol 13, pp119–203

11

A Review of ESP-r's Flexible Solution Approach and its Application to Prospective Technical Domain Developments

J. A. Clarke, N. J. Kelly and D. Tang

Abstract

This paper reviews the cooperating solver approach to building simulation as encapsulated within the ESP-r system. The application of the approach to the core technical domains underpinning building simulation is discussed along with its extension to the additional, diverse domains needed to support broader and/or more detailed analysis. The implications for computational overhead through expanding the repertoire of ESP-r are also considered. The paper concludes with an appraisal of the ability of the cooperating solver approach to cater for anticipated future application demands.

■ *Keywords* – Building performance; integrated modelling; equation solution methods

INTRODUCTION

The present drive towards sustainability in the built environment raises a number of challenges for design practitioners. These relate to the need to integrate low carbon energy sources and dramatically reduce energy consumption, while meeting increased expectations for human comfort/health and environmental protection. Detailed simulation tools may be expected to play a leading role in addressing these challenges by assisting designers to cope with complexity and differentiate design options in terms of their life cycle cost and performance.

There are two clearly identifiable tool evolution trends. First, greater modelling realism is being introduced by providing the means to model ever more complex phenomena. Second, tool scope is being expanded to include all technical domains that impact on overall performance acceptability – including, but not limited to, air quality, comfort, energy use, integrated renewable energy systems, emissions and embodied energy. In the context of the ESP-r system (Clarke, 2001), some recent innovations include modelling constructs addressing building integrated computational fluid dynamics (CFD) (Clarke et al, 1995), conjugate heat and mass transfer within building materials (Nakhi, 1995), mould growth (Rowan et al, 1997), building integrated renewables (Clarke and Strachan, 1994; Clarke et al, 1996), electrical energy systems (Kelly, 1998), fuel cells (Beausoleil-Morrison et al, 2002), stochastic occupant behaviour (Bourgeois et al, 2005), and micro-CHP (combined heat and power), (Ferguson and Kelly, 2006).

Unfortunately developments such as these portend an increased computational burden. While the adoption of more powerful computers may be acceptable within academia, the low operating margins and tight deadlines of industry will render this solution unsuitable. One possibility is to adopt a modular solution approach that can accommodate greater realism without vastly increasing the computational burden. Such an approach would allow simulation problems to be addressed at different levels of complexity depending upon the design questions in hand. This paper reviews the cooperating solution methods presently employed within the ESP-r system to solve the conservation equations relating to the interacting technical domains: building thermal processes, inter-zone airflow, intra-zone air movement, heating, ventilation and air-conditioning (HVAC) systems and electrical power flow. The paper also describes the solver adaptations required to support more rigorous domain analysis including occupant interaction, and future applications such as the modelling of cooperative micro-grids.

INTEGRATED SIMULATION

ESP-r's building model comprises a number of coupled polyhedral zones that describe geometry and fabric. Augmenting these zones is a series of networks, each of which describes an energy subsystem such as heating and air-conditioning plant, air, fluid and moisture flow. The combination of the thermal zones and associated networks, together with occupancy characteristics, user-defined control strategies and climate data, forms a complete model of the building. Clearly, useful models may start out partial and simplified and then be evolved as the design progresses.

CONTROL VOLUME MODELLING

ESP-r applies the same basic modelling principle throughout. The building is broken down into many small control volumes to which the conservation of mass, energy, momentum, species and so on can be applied. While the characteristics of control volumes can vary – homogeneous or non-homogeneous, solid or fluid, size and shape – the conservation principle can be uniformly applied as appropriate. These control volumes may be used to model a building in whole or part while including the multiplicity of interacting physical processes. Typically, a building model will comprise many thousands of control volumes, each assigned one or more conservation equations depending on the physical processes impacting on the region.

Control volume equations represent the fundamental physical processes occurring within the building, for example, heat conduction and storage, air and moisture movement, radiation exchange, electricity flow and so on. Within ESP-r, related sets of equations are grouped and each group is processed by a tailored solver that is optimized for the equation set type. Solution of the equation sets, with real climate data and user-defined control objectives as boundary conditions, gives rise to a detailed and integrated view of performance evolution over time.

A particular attraction of the control volume approach is that the same physical process can be modelled at different levels of detail. For example, the air within a zone may initially be modelled as a single control volume, which corresponds to the assumption that the air is well mixed. At a later stage this single control volume can be replaced by multiple control volumes to facilitate network airflow (Inard et al, 1996) or enhanced resolution computational fluid dynamics (Negraõ, 1995). Issues such as thermal stratification and the distribution of contaminants may then be addressed. Another example of evolving model resolution is where an initial one-dimensional (1-D) building fabric model is replaced by a 2- or 3D model to represent the effects of thermal bridges or complex constructional detailing. The further addition of moisture balance equations to the control volumes would then facilitate the modelling of coupled heat and moisture flow. The point is that it is relatively straightforward to automatically alter the number of control volumes or add additional control volume equations to reflect new inputs and analysis requests by users.

MODULAR SOLUTION

Within ESP-r, the set of equations associated with each technical domain is solved by optimized methods that are targeted to the specific nature of the equations – be they sparse, linear, non-linear, mathematically stiff and so on. The important point is that no attempt is made to concurrently solve the whole system equations. Instead, the domain equations are solved independently under the control of a supervisory routine that respects the physical couplings between domains (for example, the bi-directional coupling between air and heat flow). Examples of important couplings include: building thermal processes and natural luminance distribution; building/plant thermal processes and distributed fluid flow; building thermal processes and intra-room air movement; building distributed airflow and intra-room air movement; electrical demand and embedded power systems (renewable energy based or otherwise); and construction heat and moisture flow. The integrated solution of all domains therefore requires the coordinated application of the domain solvers applicable to the particular model being used. This approach of 'minimal adequacy' ensures that the user has access to appropriate modelling power at minimum computational effort.

BUILDING THERMAL PROCESSES

The conductive, convective and radiation exchanges associated with a building's constructions are established as a set of energy balance equations and a direct solution method applied. The approach is based on a semi-implicit scheme, which is second-order time accurate, unconditionally stable for all space and time steps, and allows time dependent and/or state variable dependent boundary conditions and coefficients. Iteration is employed for the case of non-linearity where system parameters (for example heat transfer coefficients) depend on state variables (for example temperature). An optimized numerical technique is employed to solve the system equations simultaneously, while keeping the required computation to a minimum.

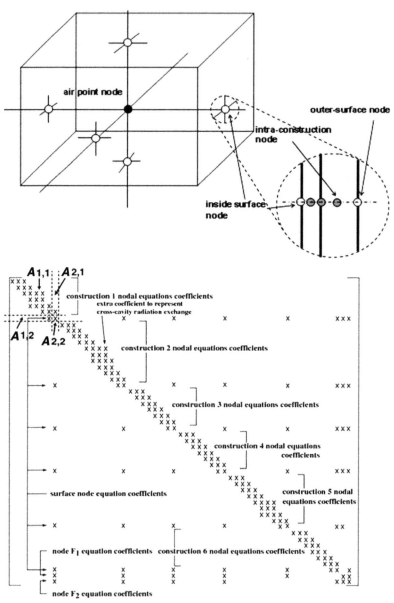

Source: chapter authors

FIGURE 11.1 The future time row matrix, *A*, for the simple room model as shown

As an example, consider the energy balance equation set for the simple room of Figure 11.1 when expressed in matrix notation: $A^{n+1} \theta = B^n \theta + C$. Here *A* and *B* are coefficients matrices corresponding to the future $(n+1)$ and present (n) time rows respectively, θ is a vector of node temperatures and flux injections, and *C* a known boundary conditions vector. Since all parameters on the right-hand side are known, this matrix equation simplifies to $A^{n+1} \theta = D$.

The form of A (Clarke, 2001) is also shown in Figure 11.1 for the case of uni-directional conduction and a single air node. The top left corner sub-matrix corresponds to the wall 1 internal nodes, while the sub-matrix (single equation) immediately below on the diagonal corresponds to the wall 1 surface node. Similarly, there are sub-matrices corresponding to constructions 2 to 6. The last coefficient on the diagonal of A corresponds to the air node within the room. The coefficients on the upper and lower off-diagonals are the radiation heat exchange coefficients connecting the inner surfaces.

Such a system of equations is solved efficiently by partitioning and reordering. Figure 11.2 shows the outcome when applied to the coefficient matrix of Figure 11.1. Note that null matrices are not shown; sub-matrices in the even rows and the last row are single equations; and θ_i, θ_{is} and θ_e correspond to the temperatures of the intra-construction, surface and air nodes respectively.

The sub-matrices may be rearranged by changing rows such that the vectors corresponding to the intra-construction nodes are moved to the upper part of the system matrix, and the vectors (single equations) corresponding to the surface nodes are moved to the lower part. This gives rise to the matrix of Figure 11.3, from which it may be observed that:

- the block matrix at the top left corner consists of sub-matrices of internal construction nodes, and is block tri-diagonal;
- the block at the lower right corner is a full block matrix, comprising surface nodes and the air node(s);
- the block matrices at the top right and lower left corners represent the connections between the innermost construction nodes and the corresponding surface node.

The lower left block matrix can be eliminated (as shown below) and thus the internal construction nodes and the surface nodes are decoupled. Therefore, only $m+1$ equations

$$
\begin{bmatrix}
A_{1,1} & A_{1,2} & & & & & & & \\
A_{2,1} & A_{2,2} & A_{2,4} & A_{2,6} & A_{2,8} & A_{2,10} & A_{2,12} & A_{2,13} \\
& & A_{3,3} & A_{3,4} & & & & \\
& & A_{4,2} & A_{4,3} & A_{4,4} & A_{4,6} & A_{4,8} & A_{4,10} & A_{4,12} & A_{4,13} \\
& & & & A_{5,5} & A_{5,6} & & & \\
& & A_{6,2} & & A_{6,4} & A_{6,5} & A_{6,6} & A_{6,8} & A_{6,10} & A_{6,12} & A_{6,13} \\
& & & & & & A_{7,7} & A_{7,8} & \\
& & A_{8,2} & & A_{8,4} & & A_{8,6} & A_{8,7} & A_{8,8} & A_{8,10} & A_{8,12} & A_{8,13} \\
& & & & & & & & A_{9,9} & A_{9,10} \\
& & A_{10,2} & & A_{10,4} & & A_{10,6} & A_{10,8} & A_{10,8} & A_{10,10} & A_{10,12} & A_{10,13} \\
& & & & & & & & & A_{11,11} & A_{11,12} \\
& & A_{12,2} & & A_{12,4} & & A_{12,6} & A_{12,8} & A_{12,10} & A_{12,11} & A_{12,12} & A_{12,13} \\
& & A_{13,2} & & A_{13,4} & & A_{13,6} & A_{13,8} & A_{13,10} & A_{13,12} & A_{13,13}
\end{bmatrix}
\times
\begin{bmatrix}
\theta_1 \\ \theta_{1s} \\ \theta_2 \\ \theta_{2s} \\ \theta_3 \\ \theta_{3s} \\ \theta_4 \\ \theta_{4s} \\ \theta_5 \\ \theta_{5s} \\ \theta_6 \\ \theta_{6s} \\ \theta_a
\end{bmatrix}
$$

Source: chapter authors

FIGURE 11.2 Partitioning of the A coefficients matrix

Source: chapter authors

FIGURE 11.3 Block partitioning of *A*

need be solved simultaneously to obtain the nodal temperatures of the wall surfaces and the air within the room (where *m* denotes the number of constructions bounding the room). Once the temperatures of the wall surfaces and the air are obtained, the temperatures of the internal construction nodes may be obtained using backward substitution. This is done by taking into account the particular features of the equation set, that is, the sub-matrices on the diagonal with odd numbers ($A_{1,1}$, $A_{3,3}$, ...) are tri-diagonal, while even numbers are single equations; most of the off-diagonal sub-matrices have only a single coefficient.

Without losing generality, the notion of matrix inversion is used. However, here the inversion requires only the elimination of the lower diagonal coefficients in the first pass. Consider the system of equations:

$$A_{1,1}\,\theta_1 + A_{1,2}\,\theta_{1s} = D_1 \tag{1a}$$

$$A_{2,1}\theta_1 + A_{2,2}\theta_{1s} + A_{2,4}\theta_{2s} + A_{2,6}\theta_{3s} + A_{2,8}\theta_{4s} + A_{2,10}\theta_{5s} + A_{2,12}\theta_{6s} + A_{2,13}\theta_a = D_{1s} \tag{1b}$$

$$A_{3,3}\theta_2 + A_{3,4}\theta_{2s} = D_2 \tag{1c}$$

$$A_{4,2}\theta_{1s} + A_{4,3}\theta_2 + A_{4,4}\theta_{2s} + A_{4,6}\theta_{3s} + A_{4,8}\theta_{4s} + A_{4,10}\theta_{5s} + A_{4,12}\theta_{6s} + A_{4,13}\theta_a = D_{2s} \quad \text{... [1d]}$$

To eliminate T_1 from the equation 1b, T_2 from the equation 1d, and so on, it is assumed that inverse matrices A^{-1} exist such that:

$$\theta_1 = A_{1,1}^{-1}\,(D_1 - A_{1,2}\,\theta_{1s}) \tag{2a}$$

$$A_{2,1}\theta_1 + A_{2,2}\theta_{1s} + A_{2,4}\theta_{2s} + A_{2,6}\theta_{3s} + A_{2,8}\theta_{4s} + A_{2,10}\theta_{5s} + A_{2,12}\theta_{6s} + A_{2,13}\theta_a = D_{1s} \tag{2b}$$

$$\theta_2 = A_{3,3}^{-1}\,(D_2 - A_{3,4}\,\theta_{2s}) \tag{2c}$$

$$A_{4,2}\theta_{1s} + A_{4,3}\theta_2 + A_{4,4}\theta_{2s} + A_{4,6}\theta_{3s} + A_{4,8}\theta_{4s} + A_{4,10}\theta_{5s} + A_{4,12}\theta_{6s} + A_{4,13}\theta_a = D_{2s} \quad \text{... [2d]}$$

Substituting the equation 2a into the 2b, the 2c into the 2d, and so on, gives:

$$Sch(A_{2,2})\theta_{1s} + A_{2,4}\theta_{2s} + A_{2,6}\theta_{3s} + A_{2,8}\theta_{4s} + A_{2,10}\theta_{5s} + A_{2,12}\theta_{6s} + A_{2,13}\theta_a = D_{1s} - A_{2,1} A_{1,1}^{-1} D_1$$

$$A_{4,2}\theta_{1s} + Sch(A_{4,4})\theta_{2s} + A_{4,6}\theta_{3s} + A_{4,8}\theta_{4s} + A_{4,10}\theta_{5s} + A_{4,12}\theta_{6s} + A_{4,13}\theta_a = D_{2s} - A_{4,3} A_{3,3}^{-1} D_2$$

$$.$$
$$.$$
$$.$$

$$\qquad\qquad\qquad\qquad\qquad\qquad\qquad\qquad\qquad\qquad\qquad\qquad\qquad\qquad (3)$$

$$A_{12,2}\theta_{1s} + A_{12,4}\theta_{2s} + A_{12,6}\theta_{3s} + A_{12,8}\theta_{4s} + A_{12,10}\theta_{5s} + Sch(A_{12,12})\theta_{6s} + A_{12,13}\theta_a = D_{6s} - A_{6,5} A_{6,6}^{-1} D_6$$

where $Sch(A_{2,2})$ is the Schur complement for $A_{2,2}$:

$$Sch(A_{2,2}) = A_{2,2} - A_{2,1} A_{1,1}^{-1} A_{1,2} \qquad\qquad (4)$$

This results in seven equations, with a full coefficient complement, to be solved simultaneously. The solution gives the temperatures of the air and surface nodes; back substituting the nodal temperatures of the surface nodes gives the temperatures of the internal construction nodes. Taken together, this procedure gives the simultaneous solution of the complete matrix equation for the room.

Considering the dimensions of $A_{1,1}$, $A_{3,3}$, and so on, and assuming each is approximately 10x10, the complete matrix A will be 67x67. Given that the method outlined above solves only seven simultaneous equations per zone, the computational saving is substantial.

The implementation of this solution procedure is complicated where intra-construction phenomena are present such as moisture transfer between the material layers, or the imposition of control-regulated heat injections/extractions corresponding to solar penetration and novel devices such as hybrid photovoltaic components and phase change materials. These phenomena require that certain terms of the corresponding conservation equations are not eliminated at matrix reduction time.

The solution process as described can be extended to any number of thermal zones, with selected parts of the model treated in greater detail. The following sections describe how the detail of different elements of a model may be increased to accommodate more complex modelling tasks.

CONSTRUCTION HEAT AND MOISTURE FLOW

To facilitate the modelling of inter-constructional moisture flow, the governing thermal equation associated with construction control volumes can be augmented with a vapour transport equation. Temperature and water vapour pressure are then both the transport potentials and the coupling variables. As implemented in ESP-r (Nakhi, 1995), water vapour flow is modelled in 1-D within a homogeneous, isotropic control volume:

$$\rho_0 \varsigma \frac{\partial (P/Ps)}{\partial t} + \frac{\partial \rho_l}{\partial t} = \frac{\partial}{\partial x}\left(\delta_P^\theta \frac{\partial P}{\partial x} + D_P^\theta \frac{\partial \theta}{\partial x}\right) + S \qquad (5)$$

where ρ is the density, o and l denote porous medium and liquid respectively, ξ the moisture storage capacity, P the partial water vapour pressure, P_s the saturated vapour pressure, δ the water vapour permeability, D the thermal diffusion coefficient and S a moisture source term. θ and P denote temperature and pressure driving potentials respectively, with the principal potential given as the subscript.

When converted to its finite volume equivalent, the above equation is non-linear and so the equations for this domain are solved by an iterative method, with linear under-relaxation employed to prevent convergence instabilities in the case of strong non-linearity or where discontinuities occur in the moisture transfer rate at the maximum relative humidity due to condensation.

AIRFLOW

ESP-r is typically used for multi-zone simulations. In this case a nodal network may be invoked to model inter-zone airflow, including infiltration and mechanical ventilation. The approach is based on the solution of the steady-state, 1-D, Navier-Stokes equation assuming mass conservation and incompressible flow. The result is a set of non-linear equations representing the conservation of mass as a function of pressure difference across flow restrictions. To solve these equations, each non-boundary node is assigned an arbitrary pressure and the connecting components' flow rates determined from a corresponding mass flow model. The nodal mass flow rate residual (error), R_i, for the current iteration is then determined from:

$$R_i = \sum_{k=1}^{N} \dot{m}_k \tag{6}$$

where $\dot{m}_k$ is the mass flow rate along the kth connection to node i, and N is the total number of connections linked to node i. These residuals are used to determine nodal pressures corrections, P^*, for application to the current pressure field, P:

$$P^* = P - C \tag{7}$$

where C is a pressure correction vector. The process, which is equivalent to a Newton-Raphson technique, iterates until convergence is achieved. C is determined from:

$$C = J^{-1}R \tag{8}$$

where R is the vector of nodal mass flow residuals, and J^{-1} is the inverse of the square Jacobian matrix whose diagonal elements are given by:

$$J_{n.n} = \sum_{i=1}^{L} \left(\frac{\partial \dot{m}}{\partial \Delta P} \right)_i \tag{9}$$

where L is the total number of connections linked to node n. This summation is equivalent to the rate of change of the node n residual with respect to the node pressure change between iterations. The off-diagonal elements of J are the rate of change of the individual component flows with respect to the change in the pressure difference across the component (at successive iterations):

$$J_{n,m} = \sum_{i=1}^{M} - \left(\frac{\partial \dot{m}}{\partial \Delta P} \right)_i ; \; n \neq m \tag{10}$$

where M is the number of connections between node n and node m.

To address the sparsity of J, its solution is achieved by LU decomposition with implicit pivoting, known as Crout's method, with partial pivoting (Press et al, 1986).

Conservation considerations applied to each node then provide the convergence criterion: $\sum \dot{m} \rightarrow 0$ at all internal nodes. As noted by Walton (1982), there may be occasional instances of low convergence with oscillating pressure corrections required at successive iterations. A relaxation factor is therefore applied using a process similar to Steffensen iteration (Conte and de Boor, 1972).

INTRA-ZONE AIR MOVEMENT

The network approach may also be used to model airflow within zones, where the zone air is represented by multiple control volumes (Inard et al, 1996). However, there are limitations to this approach. First, there is no way to account for the conservation of momentum in the flow and so this modelling method is not applicable for driving flows. Second, the inter-volume couplings are characterized as a function of pressure difference and this requires the specification of a discharge coefficient, C_d, and little research has been done to determine applicable values.

Given these limitations, ESP-r employs a built-in CFD model for intra-zone airflow simulation. Flow inside a room is characterized by a set of time-averaged conservation equations for the three spatial velocities (U, V, W), temperature θ and concentration C. For turbulent flows two additional equations are added, for turbulence intensity (k) and its rate of dissipation (ε). This is the well-known k-ε model. As with the building thermal domain, these conservation equations are discretized by the finite volume method (Negraõ, 1995; Versteeg and Malalasekera, 1995) to obtain a set of linear equations of the form:

$$a_p \phi_p = \sum_i a_i \phi_i + b \tag{11}$$

where ϕ is the relevant variable of state, p designates a cell of interest, i designates the neighbouring cells, b relates to the source terms applied at p, and a_p, a_i are the self- and cross-coupling coefficients respectively.

Because these equations are strongly coupled and highly non-linear, they are solved iteratively for a given set of boundary conditions. The SIMPLEC method is employed (Patankar, 1980; Van Doormaal and Raithby, 1984) in which the pressure of each cell is linked to the velocities connecting with surrounding cells in a manner that conserves continuity. The method accounts for the absence of an equation for pressure by establishing a modified form of the continuity equation to represent the pressure correction that would be required to ensure that the velocity components determined from the momentum equations move the solution toward continuity. This is done by using a guessed pressure field to solve the momentum equations for intermediate velocity components U, V and W and then using these velocities to estimate the required pressure field correction from the modified continuity equation. The energy equation, and any other scalar equations (for example, for species concentration), are then solved and the process iterates until convergence is attained. To avoid numerical divergence, under-relaxation is applied to the pressure correction terms.

The solution of the discretized flow equations is achieved using the tri-diagonal matrix algorithm, which is favoured because of its modest storage requirements and computational speed.

HVAC SYSTEMS

As with the zones that make up the building model, the control volume technique can also be used to derive the equation sets describing a building's plant network. Within ESP-r, the plant network consists of a coupled group of plant component models, each described by one or more control volumes. McLean (1982), Tang (1985), Hensen (1991) and Aasem (1993) have applied the control volume modelling technique to a variety of plant components. The resulting library of models allows the simulation of air-conditioning systems, hot-water heating systems and mixed air/hot water systems as a network in ESP-r. Chow et al (1998) extended the capability of the control volume approach to plant modelling by identifying 27 fundamental 'primitive parts', which are essentially elemental control volumes describing the basic thermodynamic processes occurring in air-conditioning systems. Based on this work any plant component can be described and modelled by simple primitive part combination.

The equations derived for plant are of a form similar to equations 1a to 1d, which allows their solution using an efficient, direct method. The solution of the plant matrix is dictated by control interaction, where the state of the system components is adjusted to bring about a desired control objective. The sensed node for plant control may be another plant component or node in the building model, for example, the flow rate of cold water into a chiller coil may be controlled based on the relative humidity in a thermal zone.

As a general rule, the plant-side matrix equation is substantially smaller than its building-side counterpart. For example, within ESP-r, the total number of equations for a domestic central heating system is approximately 150, while a building-side model for an average-sized house will require approximately 1000 equations. It is therefore possible to process the plant model as two equation sets for energy and mass balance (up to two phases are permitted) without the application of partitioning to accommodate sparsity. These equation sets appear as additional sub-matrices in the A matrix shown in Figure 11.2.

ELECTRICAL POWER FLOW

The approach to network airflow, as elaborated previously, can also be applied to resolve the electrical power flows associated with the building (Kelly, 1998). The electrical circuit is conceived as a network of nodes representing the junctions between conducting elements and locations where power is extracted to feed loads, or added from the supply network or embedded renewable energy components. Application of Kirchhoff's current law to some arbitrary node, i, with N connected nodes, forms the basis for the network power flow solution:

$$\sum_{j=1}^{n} \bar{I}_{i,j} = 0 \tag{12}$$

where $\bar{I}$ is a complex number describing the magnitude of the current and its phase angle. $\bar{I} = I < \phi$. The following two equations (adapted from Gross, 1986) can be derived for any node i in an AC power systems model if the impedance characteristics of the connecting electrical equipment are known:

$$\sum_{p=1}^{n} V_i V_p Y_{i,p} \cos(\theta_i - \theta_p - \alpha_{i,p}) = -\sum_{q=1}^{y} P_{G_i^q} + \sum_{r=1}^{z} P_{L_i^r} \tag{13}$$

This equation represents a real power balance associated with a node i. V is voltage, Y is admittance (the inverse of impedance Z); θ and α are the phase angles associated with V and Y respectively, P is real power (W), subscript p refers to some node connected to I, G is associated with power injected into node i, and L represents power drawn from the node.

A similar expression can be derived for the reactive power, Q (VAr), associated with node i:

$$\sum_{p=1}^{n} V_i V_p Y_{i,p} \sin(\theta_i - \theta_p - \alpha_{i,p}) = -\sum_{q=1}^{y} Q_{G_i^q} + \sum_{r=1}^{z} Q_{L_i^r}. \tag{14}$$

The solution procedure is identical to that employed for inter-zone airflow, except that here the state variable is voltage, not pressure, and two equation sets must be solved corresponding to real and reactive power flows.

SOLUTION PROCESS

The previous sections have shown how an ESP-r model may comprise many different and diverse constituent parts of varying levels of detail (for example constructions, moisture flow, electrical networks, airflow networks and so on). However, as each part is based on the same finite volume considerations, when connected together they form a consistent mathematical description of the building, no matter the level of detail adopted.

LINKING DOMAINS

The whole system problem can now be stated as the coordinated solution of the domain equations under control action that links certain model parameters (for example room air temperature to the mass flow rate induced by a fan).

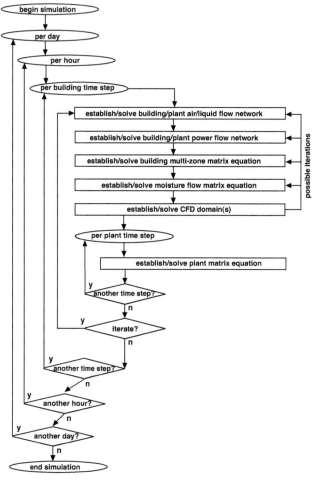

FIGURE 11.4 Iterative solution of nested domains

More specifically, the solution process employed within ESP-r links the various technical domain solution methods (direct or iterative) together through a flexible process of iterative 'handshaking' at key linkage points. This handshaking involves solving the subsystem equation sets separately, based on previous time step values of the coupled variables and then iterating to a solution. Figure 11.4 summarizes the approach.

At each building-side time step and for a given climate boundary condition, the airflow/liquid flow networks corresponding to the building and plant are established, control considerations imposed and the equations solved. Solution of these networks gives the airflow and working fluid flow rates throughout the building and within the plant system respectively.

The electrical power flow network representing building-side entities (for example lighting, small power, photovoltaic (PV) façades and so on) and plant components (for example fans, pumps, CHP plant and so on) is established, constrained by control action and the equations solved. The facility may be used to impose demand-side actions on load consuming systems. (This network model and the preceding one for airflow/liquid flow may also be invoked at higher frequency from within the HVAC solution loop.)

The building-side, multi-zone matrix equation is then established using the latest estimates of the fluid/power flows and plant-induced flux injections/extractions. Equation solution is achieved as described previously to obtain the building's temperatures and heat flows. Using the newly computed intra-construction temperatures, the construction moisture flow matrix equation is established and solved. This gives the moisture distribution within the building fabric. Using the building temperatures and airflow rates as boundary conditions, the CFD model is established and solved. This gives the intra-zone distribution of temperature, velocity, pressure and contaminants. The building temperatures and airflow/liquid flow rates are then used, along with relevant control loops, to establish and solve the plant heat and mass flow matrix equations. Solution of these equations gives the plant temperatures and flow rates.

VARIABLE FREQUENCY PROCESSING

The various equation sets for the building model represent systems with very different time constants: in the fabric of the building, temperatures change slowly over a period of hours, while in the plant system, temperatures and flow change more rapidly and certainly within minutes. A feature of the modular solution process is the ability to vary the solution frequency of each equation set, enabling the solution to capture the dynamics of a particular system. The zonal energy equations can be processed at hourly or sub-hourly intervals, while the plant and flow equations sets can be processed at a higher frequency that suits their much smaller time constants. The advantage of this process is that plant and flow can be rigorously simulated without adding to the computational overhead of solving the much larger number of building equations at an unnecessarily high frequency.

SOLUTION COORDINATION

To orchestrate the solution process, domain-aware *conflation controllers* are imposed on the different iterations. Consider, for example, the linking of the building thermal, network airflow and CFD models. This employs a controller that ensures that the CFD model is appropriately configured at each time step (Beausoleil-Morrison, 2000).

At the start of a time step, the zero-equation turbulence model of Chen and Xu (1998) is employed in investigative mode to determine the likely flow regimes at each surface (forced, buoyant, fixed, fully turbulent or weakly turbulent). This information is then used to select appropriate surface boundary conditions, while the estimated eddy viscosity distribution is used to initialize the k and ε fields. A second CFD simulation is then initiated for the same time step.

On the basis of the investigative simulation, the nature of the flow at each surface is evaluated from the local Grashof (*Gr*) and Reynolds (*Re*) numbers, which indicate *how buoyant* and *how forced* the flow is respectively:

- $Gr / Re^2 << 1$ – forced convection effects overwhelm free convection;
- $Gr / Re^2 >> 1$ – free convection effects dominate;
- $Gr / Re^2 \approx 1$ – both forced and free convection effects are significant.

Where buoyancy forces are insignificant, the buoyancy term in the *z*-momentum equation is discarded to improve solution convergence. Where free convection predominates, the log-law wall functions are replaced by the Yuan et al (1993) wall functions and constant boundary conditions imposed where the surface is vertical; otherwise a convection coefficient correlation is prescribed (this means that the thermal domain will influence the flow domain but not the reverse). Where convection is mixed, the log-law wall functions are replaced by a prescribed convection coefficient boundary condition. Where forced convection predominates, the ratio of the eddy viscosity to the molecular viscosity (μ_t / μ), as determined from the investigative simulation, is examined to determine how turbulent the flow is locally:

- $\mu_t / \mu \leq 30$ – the flow is weakly turbulent and the log-law wall functions are replaced by a prescribed convection coefficient;
- $\mu_t / \mu \leq 30$ – the log-law wall functions are retained.

The iterative solution of the flow equations is re-initiated for the current time step. For surfaces where h_c correlations are active, these are shared with the building thermal model to impose the surface heat flux on the CFD solution. Where such correlations are not active, the CFD-derived convection coefficients are inserted into the building thermal model's surface energy balance.

Where an airflow network is active, the node representing the room is removed and new connection(s) added to effect a coupling with the appropriate domain cell(s) (Clarke et al, 1995; Negrão, 1995). A technique by Denev (1995) is employed to ensure the accurate representation of both mass and momentum exchange in the situation where CFD cells and network flow components are of dissimilar size.

Similar conflation mechanisms exist to enact informed handshaking between the other domain pairings.

FUTURE REQUIREMENTS

As mentioned in the introduction, the remit of simulation tools is expanding to meet the challenges posed by sustainability and increased user expectations, among other things. This section highlights two specific examples of how the modular solution approach described here has been adapted to cope with different simulation needs, specifically the modelling of indoor air quality and micro-generation. These examples demonstrate how the approach is flexible enough to accommodate new technical domains as the need arises.

HEALTH AND COMFORT: AIRFLOW MODELLING

With regard to assessing human health and comfort the issue of indoor air quality is of high importance. Numerous issues can be addressed through the application of macroscopic and microscopic airflow modelling or a combination of the two. Examples of where airflow modelling may be deployed include (in the domestic context) prediction of condensation, dampness, mould growth and the transport of dangerous pollutants such as mould spores or carbon monoxide. In commercial and industrial contexts, airflow modelling can address some of the problems surrounding complex issues such as sick building syndrome and occupant productivity. Furthermore, the modelling of air quality can address important safety issues such as the removal of smoke and hazardous contaminants.

While the application of CFD to the above issues is not new, the combination of CFD with dynamic building simulation provides opportunities for more accurate modelling through better assessment of boundary conditions. An example of this is in the simulation of mechanically ventilated indoor car parks. The air temperature and velocity fields and contaminant concentrations can be modelled using a building model augmented with a CFD domain, while the ventilation plant can be modelled using an HVAC and associated network airflow model. The effect of supply or extract to the space, with associated dampers and diffusers, can be modelled by the imposition of pressure and velocity boundaries in the CFD domain. Conversely the HVAC system can be controlled using the spatially and temporally varying contaminant concentrations calculated by the CFD domain.

The prediction of surface condensation and mould growth is an area where the CFD and fabric moisture domains can be applied jointly. Both models can cooperate with the building thermal model to assess wall surface and near-wall conditions. Outputs required for prediction of phenomenon such as mould growth include the relative humidity and temperature at the wall surface. Using the same combined model, it would be possible to track the likely diffusion path of mould spores.

At a lesser level of granularity, a network airflow model may be used in the analysis of contaminant dispersal throughout a building. An example of this approach is the modelling of the diffusion of combustion-based contaminants through a naturally ventilated building close to a busy road. If more detail is required with regards to specific localized concentrations, the predictions of the network airflow model can be used as boundary conditions for the CFD domain. This demonstrates an advantage of the modular solution approach in that the same phenomena can be examined at very different levels of detail depending upon the context of the analysis.

LOW CARBON ENERGY SYSTEMS: MICRO-GENERATION

ESP-r has proven amenable to the modelling of passive and active renewable energy systems, with the implementation of models of solar thermal collectors (McLean, 1982), solar roof ponds (Clarke and Strachan, 1994), PV components (Clarke et al, 1996) and building integrated wind turbines (Grant and Kelly, 2004). These models are implemented either by utilizing ESP-r's existing building fabric model (for example photovoltaics) or by developing a component for use within the HVAC network model (for example solar thermal collector). Indeed, some of ESP-r's low carbon energy models span up to three domains. For example, a PV model typically comprises fabric, flow and electrical constituents. The low carbon component models are effectively the linkage points between the domains, interchanging the key coupling variables as the simulation progresses.

The HVAC domain is commonly used to model conventional HVAC systems such as variable air volume (VAV), constant air volume (CAV) and dual-duct air conditioning. The approach has also been used to model low carbon components such as fuel cells and Stirling engines (Beausoleil-Morrison et al, 2002; Ferguson and Kelly, 2006). The extensibility of the approach is essentially unlimited: new modelling methods may be implemented as new products emerge. The major issue confronting ESP-r is the generation of the component models in the first place and the combination of the selected components to form a working HVAC system. To this end, two issues need to be addressed: the synthesis of component models from basic heat transfer/flow elements, and the automatic linking of the resulting models. The former issue is addressed by the primitive part technique (Chow et al, 1998) whereby component models may be synthesized as and when required; while the latter issue might build upon previous research into object-oriented HVAC (Tang, 1996).

The ability to couple the HVAC domain to ESP-r's other domains enables it to be used in the modelling of many low carbon energy systems. For example, ventilated PV façades can be modelled using a combination of an HVAC network, airflow network and detailed model of a façade, where the PV elements are explicitly modelled as part of the building construction, interacting with the model through alteration to the fabric control volume source term (reducing the absorbed solar radiation to account for its conversion to electricity). The PV elements can themselves be coupled to an electrical network.

A potential paradigm for future energy systems is the concept of the micro-grid, where many small building-integrated power generators cooperate in a small electricity network. Building simulation tools with the appropriate domain models are ideally suited to the analysis of such systems where the production of heat and electricity is inherently linked to the time-varying building loads. Further, the approach may be extended to the modelling of communities as opposed to individual dwellings. In this regard, some challenges remain to be resolved, including the development of demand management algorithms that may be applied to switch certain loads within the context of renewable power trading. Such a facility would primarily interact with the building thermal, HVAC and electrical domains by acting to reschedule heating/cooling system set-point temperatures, and withholding/releasing power-consuming appliances where acceptable.

If more detail is required in the modelling of a specific component, it is possible to dedicate a domain to that component. For example, if a user is interested in the detailed analysis of a solid oxide fuel cell within the context of a domestic CHP unit, then a specific domain could be established to represent the fuel cell. This new domain could comprise multiple control volumes to represent the fuel cell plates, gas channels and balance of plant such as the gas desulphurizer and reformer. The equation set describing the fuel cell would include those for 3-D heat conduction in the stack, gas dynamics and electrochemical reactions. The boundary condition for these equations would come from the other domains: environment temperature, relative humidity, electrical demand and so on.

As with the CFD and network flow approaches to modelling contaminant dispersal, the ability to treat different areas of a large model at different levels of detail, while maintaining a consistent mathematical approach to the discretization of the whole physical system, is one of the key advantages of the modular approach.

SOLVER DEVELOPMENTS

The ESP-r model has proved to be resilient when applied to a range of problems over two decades. However, the real issue is whether the underlying approach will continue to be able to accommodate future user requirements. In some respects this is assured. The Open Source (www.opensource.org) nature of ESP-r should ensure that it continues to evolve in the light of new research findings. Further, the modelling of new phenomena often only requires minor changes to the existing set-up. For example, modelling of new thermal processes occurring in the building fabric requires only the implementation of a new source term, the adjustment of existing equation coefficients or incorporating both together through a control function. This approach has been adopted in recent work that focused on the implementation of phase change materials, while ongoing work is addressing the nuances of double skin façade modelling. In both cases, no solver adaptations are required. The system has also proved suited to the modelling of innovative components such as light-sensitive shading devices and hybrid PVs, both of which required enhancements to the resolution of existing models to allow, for example, slat angle adjustment in the former case and heat transfer surface geometry modelling in the latter. Other developments to the solver may require the creation of an entirely new domain model, as in the case of detailed fuel cell modelling set out above. The issue then becomes the ease with which the new domain can be integrated.

Referring to the requirements for health and sustainability set out above, specific solver developments could include extension of construction moisture flow modelling from 1-D to 3-D, to facilitate more accurate representation of moisture transport within building surfaces. This would be an important addition to ESP-r's capabilities in the modelling of surface phenomenon such as condensation and growth of biological contaminants. It would not be appropriate to couple a detailed surface model to a lumped air volume as the detail of the surface model predictions would be lost. Rather, adequate representation of surface and air vapour distribution dictates that a CFD domain couples to the detailed surface model. Such an extension impacts upon the treatment of surface (de)absorption in the presence of an active CFD domain. This will require extensions to the linkages between the building thermal/moisture and CFD domains to handle the possible gridding cases: 1-D/3-D, 2-D/3-D and 3-D/3-D.

With regards to CFD, this is already widely applied to problems in building airflows, and the integrated CFD domain solver within ESP-r can be readily applied to many air quality issues with minor adaptation.

Other developments will require additional equations and so will impact upon the solution procedure. For example, the CFD technique is presently incapable of modelling small solid particulates that have appreciable mass within the main airstream. Two particle dispersion/deposition modelling methods are being considered for implementation: treating the particles as a continuum, or particle tracing using Lagrange coordinate.

A prototype continuum-based model has been recently implemented to enable the modelling of very small particles in flows with Stoke's number significantly less than 1 (Kelly and Macdonald, 2004).

The modelling of fire/smoke requires that extra equations be added to handle combustion reactions and the transport of combustion products (for example via the implementation of mixture fraction or grey gas radiation transport models). Such adjustments can be readily implemented within the existing code since the governing equations are of diffusive type and so can be treated in the same way as the energy and species diffusion equations. In conjunction with the network flow model, the transient distribution of fire and smoke may then be applied as a boundary condition for the prediction of the movement of occupants during a fire.

In relation to airflow modelling using the network approach, possible improvements include the introduction of a pressure capacity term into the mass balance equations to facilitate the modelling of compressible flows and the introduction of transport delay terms to the network connections. Indeed, if the fluid velocity and geometrical characteristics of a particular connection are known, then a transport delay can be calculated automatically.

Further developments are also required in relation to the connection of electrical domains, especially in relation to the modelling of many of the micro-generation systems described above. These will often be subjected to control actions based on electrical as opposed to thermal criteria. Examples include voltage regulation, network stability and phase balancing. Control of this type requires the creation of high-level controllers to iteratively couple the electrical and other associated domains. Such controllers will need to be intelligent enough to balance the conflicting demands of local comfort and community benefit within the micro-grid. This type of control is likely to include some form some of financial decision-making.

Finally, the detailed, dynamic modelling of micro-grids will require the simultaneous modelling of many different buildings and micro-generation systems along with the electrical system to which they are connected. Modelling at this scale would appear to be well suited to parallel processing, where different aspects of the model (for example dwellings, micro-power systems and the electrical grid) can be allocated to different streams.

In the long term, additional solver developments may be implemented to bring about computational efficiencies and thereby assist with the translation of simulation to the early design stage. For example:

- Additional, context-aware solution accelerators may be embedded within the solvers to control their appropriate invocation.
- Parallelism may be introduced to allow the different domains to be established and solved in tandem to reduce simulation times.
- Network computing might be exploited to allow different aspects of the same problem to be pursued at different locations as an aid to team working.

Such developments might well be built upon entirely new methods such as 'intelligent matrix patching', whereby the coupling information between domain models are stored in a 'patch matrix' allowing the numerical model of the coupling components to be activated only when the actual coupling takes place. Furthermore, a greater level of coefficients management may be introduced to ensure that the matrix coefficients are only updated when required and otherwise never reprocessed. Such devices would lead to significant reductions in computing times.

Finally, it is worth mentioning that no matter how well the technical domains of the building are modelled, the effects of occupancy can vastly alter the physical behaviour and ultimately the predicted energy performance. These effects arise from two avenues: behaviour (for example the occupants' response to window opening) and attitude (for example the rejection of facilities on grounds other than performance). In an attempt to facilitate the modelling of occupancy on the various technical domains that comprise an ESP-r model, two approaches are possible: 'typical' interactions may be included within a controller that has authority to adapt the parameters of the affected domain models prior to solution; or, more realistically, an occupancy response model may be introduced by which the response to stimulus is explicitly represented. In both cases the aim is to address the distributed impacts of occupant actions (for example the impact on the network flow, CFD, thermal and lighting domains of window opening).

CONCLUSIONS

This paper has summarized the modular approach to multi-domain solution as employed within the ESP-r system. Several possible technical developments were then outlined as required by new user demands that imply a need to extend and deepen future building performance analysis. The modular approach was shown to be well suited to future adaptation in response to these new demands.

AUTHOR CONTACT DETAILS

J. A. Clarke, Energy Systems Research Unit, University of Strathclyde, Glasgow, G1 1XJ, UK

N. J. Kelly, Energy Systems Research Unit, University of Strathclyde, Glasgow, G1 1XJ, UK

D. Tang, Shanghai Pacific Energy Centre, Suite 901, Info Tower, 1555 Kongjiang Road, Shanghai 200092, China

ACKNOWLEDGEMENTS

The authors are indebted to the many individuals who have contributed to the ESP-r project since its inception in 1974.

REFERENCES

Aasem, E. O. (1993) *Practical Simulation of Buildings and Air-conditioning Systems in the Transient Domain*, PhD thesis, University of Strathclyde, Glasgow

Beausoleil-Morrison, I. (2000) *The Adaptive Coupling of Heat and Air Flow Modelling within Dynamic Whole-Building Simulation*, PhD thesis, University of Strathclyde, Glasgow

Beausoleil-Morrison, I., Cuthbert, D., Deuchars, G. and McAlary, G. (2002) 'The Simulation of Fuel Cell Cogeneration Systems in Residential Buildings', *Proceedings of eSim, Bi-annual Conference*, IBPSA-Canada

Bourgeois, D., Reinhart, C. F. and Macdonald, I. A. (2005) 'Adding advanced behavioural models in whole building energy simulation: A study on the total energy impact of manual and automated lighting control', *Energy and Buildings*, vol 38, no 7, pp814–823

Chen, Q. and Xu, W. (1998) 'A zero-equation turbulence model for indoor airflow simulation', *Energy and Buildings*, vol 28, no 2, pp137–44

Chow, T. T., Clarke, J. A. and Dunn, A. (1998) 'Theoretical basis of primitive part modelling', *ASHRAE Transactions*, vol 104, no 2, pp299–312

Clarke, J. A. (2001) *Energy Simulation in Building Design*, 2nd edition, Butterworth-Heinneman, London

Clarke, J. A. and Strachan, P. A. (1994) 'Simulation of conventional and renewable building energy systems', *Proceedings of 3rd World Renewable Energy Congress*, Part II, Reading

Clarke, J. A., Dempster, W. M. and Negrão, C. (1995) 'The implementation of a computational fluid dynamics algorithm within the ESP-r System', *Proceedings of Building Simulation 1995*, Madison, WI

Clarke, J. A., Hand, J. W., Johnstone, C., Kelly, N. and Strachan, P. A. (1996) 'The characterisation of photovoltaic-integrated building façades under realistic operating conditions', *Proceedings of the 4th European Conference on Solar Energy in Architecture and Urban Planning*, Berlin

Conte, S. D. and de Boor, C. (1972) *Elementary Numerical Analysis: An Algorithmic Approach*, McGraw-Hill, New York

Denev, J. A. (1995) 'Boundary conditions related to near-inlet regions and furniture in ventilated rooms', *Proceedings of The Application of Mathematics in Engineering and Business*, Technical University of Sofia, Sofia, pp243–248

Ferguson, A. and Kelly, N. J. (2006) 'Modelling building-integrated Stirling CHP systems', *Proceedings of eSim, Bi-annual Conference*, IBPSA-Canada

Grant, A. D. and Kelly, N. J. (2004) 'A ducted wind turbine model for building simulation', *Building Services Engineering Research and Technology*, vol 25, no 4, pp339–350

Gross, C. A. (1986) *Power System Analysis*, 2nd edition, John Wiley, New York

Hensen, J. L. M. (1991) *On the Thermal Interaction of Building Structure and Heating and Ventilating System*, PhD thesis, University of Technology, Eindhoven

Inard, C., Bouia, H. and Dalicieux, P. (1996) 'Prediction of air temperature distribution in buildings with a zonal model', *Energy and Buildings*, vol 24, pp125–132

Kelly, N. J. (1998) *Toward a Design Environment for Building-Integrated Energy Systems: The Integration of Electrical Power Flow Modelling with Building Simulation*, PhD thesis, University of Strathclyde, Glasgow

Kelly, N. J. and Macdonald, I. (2004) 'Coupling CFD and visualisation to model the behaviour and effect on visibility of small particles in air', *Proceedings of eSim Bi-annual Conference*, Vancouver

McLean, D. J. (1982) *The Simulation of Solar Energy Systems*, PhD thesis, University of Strathclyde, Glasgow

Nakhi, A. E. (1995) *Adaptive Construction Modelling Within Whole Building Dynamic Simulation*, PhD thesis, University of Strathclyde, Glasgow

Negrão, C. O. R. (1995) *Conflation of Computational Fluid Dynamics and Building Thermal Simulation*, PhD thesis, University of Strathclyde, Glasgow

Patankar, S. V. (1980) *Numerical Heat Transfer and Fluid Flow*, Hemisphere, New York

Press, W. H., Flannery, B. P., Teukolsky, S. A. and Vettlering, W. T. (1986) *Numerical Recipes: The Art of Scientific Computing*, Cambridge University Press, Cambridge

Rowan, N. J., Anderson, J. G., Smith, J. E., Clarke, J. A., McLean, R. C., Kelly, N. J. and Johnstone, C. M. (1997) 'Development of a computer program for the prediction of mould growth in buildings using the ESP-r modelling system', *Indoor Built Environment*, vol 6, pp4–11

Tang, D. (1985) *Modelling of Heating and Air Conditioning System*, PhD thesis, University of Strathclyde, Glasgow

Tang, D. (1996) 'Object technology in building environmental modelling', *Building and Environment*, vol 32, no 1, pp45–50

Van Doormaal, J. P. and Raithby, G. D. (1984) 'Enhancements of the SIMPLE method for predicting incompressible fluid flows', *Numerical Heat Transfer*, vol 7, pp147–63

Versteeg, H. K. and Malalasekera, W. (1995) *An introduction to Computational Fluid Dynamics: The Finite Volume Method*, Longman, Harlow

Walton, G. N. (1982) 'Airflow and multiroom thermal analysis', *ASHRAE Transactions*, vol 88, no 2, pp78–91

Yuan, X., Moser, A. and Suter, P. (1993) 'Wall functions for numerical simulation of turbulent natural convection along vertical plates', *International Journal of Heat Mass Transfer*, vol 36, no 18, pp4477–4485

Index

Milton Keynes UK
Ingram Content Group UK Ltd.
UKHW040108071024
449327UK00019B/911

9 780367 577704